Design

At the end of the First World War Germany was humbled. Gone was the Hohenzollern dynasty to be replaced by the Weimer Republic. The determination of the allies, especially France, to take revenge on Germany proved short-sighted in the extreme, as all this achieved was to make the struggling German republic a breeding ground for seething resentment. The end product was Adolf Hitler. Imperial Germany had been a latecomer as a major naval power and the explosive growth of the Imperial German Navy had been one of the causes of the war. After the Armistice the High Seas Fleet was supposed to be interned in a neutral port pending a peace treaty, but this port turned out to be Scapa Flow, home of the Grand Fleet. While the British Fleet was at sea on manoeuvres most of the once-proud capital ships of the High Seas Fleet scuttled themselves rather endure a more shameful fate. The eventual peace treaty signed at Versailles included clauses restricting the development of any future German navy. As foreseen, all modern German warships were parcelled out among the victors and Germany was only left with a handful of obsolete predreadnought battleships.

Future construction was hamstrung by size constraints. Maximum displacement was set at 6000 tons for cruisers and at 10,000 tons for battleships. The Germans were allowed to replace their old predreadnought battleships on a one-for-one basis as each old battleship reached 20 years of age. Great Britain and France could not foresee Germany producing anything that could be considered a threat on a 10,000-ton displacement, but German naval architects took up this challenge and, by thinking outside of the box, came up with a unique answer. This was not a conventional battleship design but instead combined the armour of a cruiser with six 11in guns. Rated simply as a 'panzerschiff' (armoured ship), the first of a three-ship class was the *Deutschland*. Arguably the most efficient German warship design, it provided a very powerful warship on a cruiser's displacement. However, the Weimer Republic lied about the displacement from the start, as the final displacement was substantially over 10,000 tons. With extremely efficient diesel engines providing a top speed of 26 knots, the only ships with superior armament and the speed to run down the panzerschiffe were the three British battlecruisers. The diesel power plant produced a huge operational range of 10,000nm, making the ships the perfect commerce raider.

The British and French admiralties were totally confounded by the appearance of the panzerschiffe. Great Britain had no battleship tonnage left under the Washington Treaty for a reply, but France did and built the *Strasbourg* and *Dunkerque* as an answer to the panzerschiffe. The last true battlecruiser design, the pair did have the speed, armament and armour to defeat a panzerschiff. Germany had already built or

Below: *Gneisenau* on 22 August 1938 off Kiel for the celebrations surrounding the launch of the cruiser *Prinz Eugen*. At this time the ship was carrying out trials with various prototype floatplanes, the aircraft on the catapult being a Heinkel He 114. Note that as completed the ship carried a stump foremast stepped on the main director on the foretop.

Above: The newly completed *Scharnhorst* in April 1939. At this stage of her career, the ship looked very like her sister, with straight stem, mainmast on the funnel platform and no funnel cap. However, the ship carried her foremast stepped on the rear of the admiral's platform from commissioning.

was building three of the type – *Deutschland*, *Admiral Scheer* and *Admiral Graf Spee* – and had plans for two more. The revelation of the French *Strasbourg* design, however, prompted a rethinking and the next two were to be improved panzerschiffe with far greater displacement. Although this design still carried six 11in guns, the armour protection and size greatly increased. With a design displacement of around 20,000 tons and a length of almost 700ft, it would have been impossible to claim that these ships were within the 10,000-ton Versailles limits. At the same time designers and admirals flirted with ways of providing heavier armament than the six 11in guns of their predecessors. One proposal was to provide two quadruple 11in guns along the lines of the mountings in the French battlecruisers; others were to match the 330mm (13in) guns of the French ships, or to mount 350mm (14in) and 380mm (15in) main armament. With any of these choices, a total redesign would be needed, substantially extending the design and construction process.

Even with this large expansion of the panzerschiff design, this pair would still have been inferior to the new French battlecruisers without the total redesign required for heavier main gun armament. Hitler had been in power since 1933 and had approved the orders for the two improved panzerschiffe. The contract for 'Panzerschiffe D' (*Ersatz Elsass*), later *Scharnhorst*, was awarded to Marinewerft, Wilhelmshaven on 24 January 1934 as a six 11in gun improved *Deutschland*. The same was true with ship 'E' (*Ersatz Hessen*), later *Gneisenau*, ordered from Deutsche Werke, Kiel on 25 January 1934. The keels for both ships were laid down on 14 February 1934 but it was only a few months before work was stopped on these hulls. Further thought had gone into the design and it was finally decided that indeed something more

Right: The ships carried nine 11in guns (formally called 28cm, but actually 283mm) SK C34 in triple Drhl C34 turrets that were virtually identical to those mounted in the panzerschiffe. This view shows *Scharnhorst*'s 'A' and 'B' turrets (known in German naval parlance as Anton and Bruno – 'C' was Caesar). On pre-war German warships the turrets were also given individual names celebrating naval tradition, in this case 'Graf Spee' and 'Gneisenau', and the cast shields that can be seen beneath the rangefinder housings carried the coats of arms of those officers' families.

DESIGN CHARACTERISTICS

Dimensions *As Built*:	Length 754ft (229.8m) oa; 741ft 5in (226m) wl. Beam 98ft 5in (30m). Draught 27ft (8.23m) mean (designed); 32ft 7in (9.93m) maximum
With Atlantik stem:	Length 772ft 4in (235.4m) oa
Displacement	26,000 tons standard (official) 34,841 tons normal load 38,900 tons full load
Armament	Nine 11in (280mm) guns (3x3); twelve 5.9in (150mm) guns (4x2,4x1); fourteen 4.1in (105mm) heavy AA guns (7x2); sixteen 37mm medium AA guns (8x2); eight single 20mm light AA guns (8x1); six 21in (533mm) torpedo tubes (3x2) added 1941/42
Vertical Armour	Main belt 320-170mm; conning tower 450-200mm; main turrets 360-150mm
Horizontal Armour	Main armoured deck 150mm; upper armoured deck 50mm
Machinery	Three high pressure turbines; twelve small water tube boilers, 160,000shp maximum
Maximum Speed	31.65kts *Scharnhorst*; 30.7kts *Gneisenau*

ARMAMENT CHARACTERISTICS

Ordnance	Purpose	Horizontal Range	Elevation/Vertical Range	Rate of Fire (rpm)	Shell Weight
11in (280mm) / L54	Surface fire	46,748yds; 42,500m	NA	3 to 4	739lbs; 336kg
5.9in (150mm) /L55	Surface fire	25,153yds; 23,000m	NA	8	100lbs; 45.3kg
4.1in (105mm) /L65	DP but primarily heavy AA	19,357yds; 17,700m	85 degrees/12,500m	15	33.3lbs; 15.1kg
37mm (1.46in) /L83 SK C30	Medium AA	7382yds; 6750m effective	80 degrees/6800m maximum/ 4800m effective	160 maximum 60 operational	1.64lbs; 0.745kg
20mm (0.79in)	Light AA	5249yds; 4800m	90 degrees/4800m maximum/ 3700m effective	280 maximum 120 operational	0.291lb; 0.132kg
21in (533mm) torpedo tubes	Sub-surface	6000m at 44kts 8000m at 40kts 14,000m at 30kts	NA	NA	3384lbs; 1538kg

powerful than a ship with six 11in guns was needed to meet the French pair. Hitler had been in negotiations with Great Britain and the result was the Anglo-German naval agreement of 1935 in which Germany was allowed to build up to 35 per cent of the Royal Navy's tonnage by type. By 1934 the outcome looked promising so on 5 July 1934 work stopped and the keels were scrapped. The contracts were not cancelled; instead a new larger design would be built.

The first of the new construction was to be much larger than the improved panzerschiffe design but, so as not to antagonize the British, it was still to carry 11in guns. Perhaps more important to the decision was the fact that of all systems which determined the building time of a battleship, armament took the longest to develop. To use any heavier gun would have added significantly to the building period. The German navy had been developing a new 15in gun but it was far from ready for deployment, so it would be far quicker to use the same armament as the *Deutschland* class, the triple 11in gun turret for the new

Scharnhorst on 20 April 1939 dressed overall to celebrate Adolf Hitler's birthday.

Above: *Gneisenau*'s forecastle and forward turrets from a bridge platform, spring 1940. Note the periscope sights on the roof of the conning tower, and the open navigation bridge around it. These ships had extending bridge wings to facilitate close-quarter manoeuvring that were hinged back against the superstructure when at sea.

Above, right: Even as refitted, *Gneisenau*'s anchors continued to be a serious source of spray at sea and eventually they were moved to sloping beds on the deck edge and the traditional hawse pipes were plated over. During this process the two portside anchors were reduced to one. This photo from sea level shows the new configuration in *Gneisenau* in the summer of 1939.

Below: Bridge and forward turrets of *Scharnhorst*, probably about November 1939. At this point, the ship still has not had the FuMO 22 radar added.

design, but with three turrets instead of two. Furthermore, the turrets were already on order for the fourth and fifth of the panzerschiffe. There was still support for 15in guns for the pair but as a compromise the barbettes for the triple 11in mounting were to be of the same dimensions as the new twin 15in turret design, allowing subsequent replacement. Accordingly, the new design was a political and practical compromise, driven by the need for speedy construction.

The decision to continue carrying single-purpose 150mm (5.9in) secondary guns for surface action instead of dual-purpose surface/AA guns can be criticized. It is true that the USN was developing DP 5in guns for their new battleship designs and the Royal Navy 5.25in guns for their new (*King George V*) design, but the Japanese clung to surface action 6.1in guns for the *Yamato* and the Italians and French also kept single-purpose secondaries for their new designs. The German 4.1in DP gun would have been considered too light in the surface role for battleship secondary armament by everyone but the few visionaries that foresaw the dominance of airpower in future conflicts. The German navy had very few, if any, of such visionaries on its staff. There certainly was no time to design a completely new dual-purpose gun heavier than the 4.1in.

The two ships were named *Scharnhorst* and *Gneisenau*, perpetuating the names of Admiral von Spee's two armoured cruisers of the First World War. Many references call these ships battlecruisers because of their light 11in gun armament but they were officially called 'schlachtschiffe' (battleships) and also continued German battleship design preferences from the Great World. Battleships of the Imperial German Navy mounted lighter guns than battleships of the Royal Navy but had much superior protec-

RECONNAISSANCE FLOATPLANE CHARACTERISTICS

	Max Speed	Ceiling	Range	Powerplant	Armament
Arado Ar 196 A3	193mph	23,000ft	670 miles	970hp BMW 132k nine-cylinder radial	Two 20mm cannon in wings; three 7.9mm MG17 machine guns, one forward and two for rear gunner; two 110lb bombs
Heinkel He 60	149mph	16,400ft	513 miles	660hp BMW V1 6.0 twelve-cylinder liquid cooled	One 7.9mm MG in dorsal position

Left: The *Scharnhorst* benefited from the experience gained with her sister and between June and August 1939 was similarly modified forward, but with the addition of a stemhead anchor, prefiguring the system adopted in *Bismarck* and *Tirpitz*. However, the refit was more radical and included the installation of a new tripod mainmast aft of the hangar, and thereafter the sisters were easily distinguishable. Note that the funnel cap was also less raked.

tion; and with the new *Scharnhorst* design, they were also much faster. The designers' first choice was diesel engines but they could not provide the desired maximum speed of 31kts. As a consequence, it was decided to use a newly designed high-pressure steam plant that would provide the power for high speed but resulted in a greatly reduced radius of action compared to the diesel-powered panzerschiffe. This high-pressure power plant never met expectations and proved to be a weakness in all heavy ships in which it was installed, such as the *Hipper* class, in that it tended to break down frequently. The power plant developed 125,000shp and *Scharnhorst* was slightly faster than her sister: she clocked in at 31.65kts on trials while *Gneisenau* reached 30.7kts. Range was less than expected at 7100nm for *Scharnhorst* (6200nm for *Gneisenau*) instead of the designed 8200nm at 19kts. For public purposes the displacement said to be 26,000 tons but, as was the case with other German designs, actual displacement was far greater. *Scharnhorst* had a standard displacement of 34,780 tons and deep load (operational) displacement of 41,757 tons. As designed and built, the pair followed the same 'family' appearance as the panzerschiffe, with an almost straight cutwater, tower superstructure and single flat-topped funnel.

The pair were 754ft (229.8m) long with their original straight stem and added another 16ft when fitted with the overhanging so-called Atlantik bow. They had a beam of 98ft 5in (30m) and draught of 27ft (8.23m). The main armour belt was 13.8in tapering to 7.9in (350-200mm). Maximum turret armour was 14.2in, 9in thicker than the *Deutschland* turrets. Barbette maximum armour was 13.8in, as was conning tower armour. The *Gneisenau* was laid down on 6 May 1935 followed by *Scharnhorst* on 15 June 1935, but *Scharnhorst* was faster in construction and was launched 3 October 1936 by the widow of the captain of the First World War *Scharnhorst*, with Hitler present, while *Gneisenau* was launched on 8 December 1936 by the widow of the captain of the armoured cruiser *Gneisenau*. However, *Gneisenau* commissioned first, on 28 May 1938 followed by *Scharnhorst* on 7 January 1939. The fitting-out schedule for *Scharnhorst* was slowed by increasing demands on the yard's capacity, as the quantity of all types of warships was expanding rapidly.

Left: *Scharnhorst* in September 1939. Two of the Arado Ar 196 floatplanes, then newly entering service, can be seen on the catapults. The aircraft on 'C' turret was clearly very vulnerable to blast and weather damage and that catapult was soon removed from both ships.

Careers

The earliest sea trials with *Gneisenau* proved that the ship was very wet forward, with spray often making it impossible to use the rangefinders in 'A' and 'B' turrets and drenching the bridge. The ship was quickly returned to Deutsche Werke and fitted with an elongated bow with increased flare, but this proved less effective than hoped and the ship was further modified by having the anchors – a major source of spray – moved from their conventional hawse holes up to the deck edge. This so-called Atlantik bow was later fitted to *Scharnhorst*, but her refit was more extensive, with the mainmast being moved from the funnel platform to the after superstructure. Both ships were also given funnel caps.

After receiving their Atlantik bows both sister ships were sent to the Baltic for a new work-up period until November 1939. Under the overall command of Admiral Marschall *Scharnhorst* and *Gneisenau* sortied on 21 November 1939. Their mission was to provide a diversion for the panzerschiffe by being seen and unsettling British dispositions. As they reached the Faroes–Iceland gap, they encountered the British patrol line in the form of the armed merchant cruiser *Rawalpindi*, a former P & O liner mounting seven 6in guns. The liner had no chance against the two battleships but sent out the alert and fought gamely until overcome. The other two members of the patrol line, HMS *Newcastle* and HMS *Cairo* came steaming to the rescue. The *Rawalpindi* was dead in the water with all guns silenced by 16:15 but did not sink until 20:00. Twenty of the AMC's crew had been rescued by the German ships when the British cruisers arrived, and although Marschall could have easily handled these two as well, he instead broke off to the

■ *SCHARNHORST* CLASS KEY DATES

	Laid Down	Launched	Commissioned	Fate
Scharnhorst	15 June 1935*	3 October 1936	7 January 1939	Sunk, 26 December 1943
Gneisenau	6 May 1935*	8 December 1936	21 May 938	Scuttled, 27 March 1945

* Both ships were first laid down on 14 February 1934, but suspended when the design was revised; work resumed to the definitive design on the dates quoted above.

Below: The ensign is hoisted for the first time at *Scharnhorst*'s commissioning ceremony, 7 January 1939. The ship's first captain, Otto Ciliax, addresses the ship's company from the top of 'C' turret. The gilded bronze eagle adorned the stern of most major German warships (and the bridge front of destroyers) during the Nazi era, but they were generally removed in the early years of the war. That from *Graf Spee* was recovered from the scuttled wreck and is now in the Montevideo maritime museum.

Below, right: *Gneisenau* on 21 November 1939 setting out on the operation against British convoys which was to result in the sinking of the AMC *Rawalpindi*.

Left: *Scharnhorst* hove to in order to recover an early model Ar 196. The ship now had FuMO 22 radar on the foretop director, so the photo probably dates from very late in 1939 or early 1940..

northeast into the Norwegian Sea. The appearance of the German pair certainly turned over the RN apple cart. *Rodney* and *Nelson* set off in one direction; *Hood* with *Dunkerque* and two French cruisers made for Iceland; *Repulse* and *Furious* rushed eastward from Canada; and *Warspite* went to the Denmark Strait. The *Scharnhorst* and *Gneisenau* made it back to Wilhelmshaven on 28 November while the British and French forces were still searching into early December. The ships were in the dockyard for December and into January before being sent to the Baltic for training in January and early February.

OPERATION WESERÜBUNG

Operation Nordmark took place during 18-20 February 1940, when both ships, along with the heavy cruiser *Admiral Hipper*, went as far as the Shetland Islands, but no contact was made and therefore nothing was achieved. The next operation would

Left: Photographed just prior to Operation Weserübung (the invasion of Norway) in April 1940, *Scharnhorst* is seen with the prominent recognition markings adopted in an attempt to deflect the unwanted attention of the Luftwaffe, who had recently been responsible for the sinking of two German destroyers. These markings included large swastikas fore and aft and coloured turret tops (usually red but sometimes yellow).

have different results: Operation Weserübung was the invasion of Denmark and Norway in April and the entire surface fleet was employed. The mission of *Scharnhorst* and *Gneisenau* was to provide cover by steaming west of Norway to screen and support the invasion forces. Before the ships left port, Admiral Marschall was replaced by Admiral Gunther Lütjens with his flag aboard *Gneisenau* (Lütjens would later command *Bismarck* and *Prinz Eugen* and famously went down with his flagship). The pair ran into a gale and had to reduce speed to 9kts, but during the storm *Scharnhorst* sustained weather damage. On 8 April destroyers carrying troops sighted a single British destroyer and Lütjens ordered *Hipper* to that position. The British ship was HMS *Glowworm*, which had been an escort for the battlecruiser *Renown* until she went back to search for a crewman who had been washed overboard in the gale. The next day the battleships encountered the *Renown* herself and nine destroyers. At 05:05 a large ship was made out in the darkness to the west. Two minutes later *Renown* opened fire but it was another three minutes before the enemy was identified as *Renown* and the German ships returned fire. On *Scharnhorst* 'A' turret could not be used because of flooding caused by the gale, and to further complicate the issue, the centreline engine stopped, the first appearance of the weakness of the power plant design that would plague the class. After 48 minutes the centreline engine came back on line but only two minutes later the port engine began to slow down; additionally two boilers went out of service. The engagement was sporadic and at long range and the ships would lose sight of *Renown* from time to time. By 07:00 the action was over but because of the power plant problems the maximum speed of *Scharnhorst* had been reduced to 18kts. During the action *Gneisenau* and *Renown* had each hit one another, but *Scharnhorst*'s gunnery was off her usual high standards. By 12 April the ships were back in port and *Scharnhorst* received a quick refit, which lasted until 10 May. The pair went back into a training cycle until June.

Since the April invasion of Norway, the German force assigned to Narvik had become hard pressed by the allies; Operation Juno was designed to relieve this pressure, and lasted from 4 to 10 June 1940. *Scharnhorst*, *Gneisenau*, *Hipper* and four destroyers were sent back to Norway with Admiral Marschall once more in command. What the Germans did not know was that the allies were already removing their troops from Norway. The Wehrmacht had unleashed the Blitzkrieg in the west and the troops were now desperately needed in France. The allied withdrawal was well under way when the German ships sailed into the middle of the operation. On 8 June the Germans came upon part of the

Below: *Gneisenau* steaming ahead of *Scharnhorst* during Operation Juno. Note that the twin 5.9in (150mm) mounting in the foreground also has its turret top painted in a dark colour like the main armament.

allied support forces. The large 19,500-ton liner *Orama*, serving as a troopship but without troops aboard, was sunk, as well as the armed trawler HMS *Juniper* and tanker *Oil Pioneer*, with *Hipper* taking on most of the action in order to save the 11in shells on the battleships. *Hipper* and the destroyers were then detached to Trondheim for refuelling. Because of this, they missed the main event of that day. *Scharnhorst* and *Gneisenau* steamed on alone after the merchant massacre and at 16:46 sighted smoke. The pair had stumbled across the carrier HMS *Glorious* with escorting destroyers *Ardent* and *Acasta*.

The *Glorious* was commanded by a martinet with no background in naval aviation who had taken umbrage at operational guidance offered by his air officer and other senior ship's officers. He was in a hurry to get back Scapa Flow to court-martial those officers and had been given permission to detach from the main force. *Glorious* also was carrying RAF aircraft and crews evacuated from Norway. Incredibly, although steaming through waters in which the Germans had operated previously, no aircraft were in the air for scouting, nor were aircraft spotted for immediate take-off. The Germans saw that they had caught a carrier and two destroyers and closed at high speed. Marschall ordered the secondary guns to engage the destroyers while the 11in guns were reserved for *Glorious*, which was at first thought to be *Ark Royal*. At 17:32 the 11in guns fired their first rounds at a range of 26km. The first hit was six minutes later. The *Glorious* never got one aircraft airborne and within an hour she was burning and listing heavily. At 18:19 *Scharnhorst* had to reduce speed because of her reoccurring nemesis, engine trouble. *Glorious* sank at 18:20, followed by *Ardent* two minutes later, but at 18:33 *Acasta* fired four torpedoes at

Above, left: *Scharnhorst* opens fire at the carrier *Glorious* on 8 June 1940. Note the air identification swastika on the forecastle (black in a white circle on a red panel), and dark painted turret tops, possibly the shade of red called *Kaiserrot* or the dark grey *Dunkelgrau*.

Above, right: A propaganda photo purporting to show *Gneisenau* engaging HMS *Renown* on 9 April 1940. Given the low angle of elevation of the guns and the relatively calm sea, this is inherently unlikely. The turret tops are in dark paint and the swastika marking on the forecastle seems to have been painted over.

Left: Trondheim Roads on 11 June 1940 occupied by German forces after the invasion of Norway. *Gneisenau* is to the left with the *Admiral Hipper* in the right background. There are six Heinkel He 115 twin-engined floatplanes in the foreground.

Scharnhorst, one of which hit her quarter in the vicinity of 'C' turret and the magazine was flooded. The starboard engine stopped and speed started dropping quickly. By 19:06 the centreline engine had also stopped, but the crew had been able to get the starboard engine running again and maximum speed went from 21kts to 24kts. With *Scharnhorst* damaged and all three British ships sunk, the battleships made for Trondheim, which was reached at around 13:00 on 9 June ; after temporary repairs they left for Kiel on 20 June. However, *Gneisenau* was hit by a torpedo from the submarine *Clyde* that blew a large hole in the bow just aft of the stem. She had to return to Trondheim for emergency repairs and it was a month before the ship could sail for Kiel. Despite the score of one carrier, two destroyers, a liner, oil tanker and trawler, the High Command was disappointed with the outcome and blamed Marschall, who was again replaced by Lütjens.

OPERATION BERLIN

The repairs and work-up to both ships lasted until December, and on the 28th the two battleships attempted a break-out into the North Atlantic, but storm damage caused them to abort the mission and return to the Baltic. On 22 January 1941 the battleships left Kiel on Operation Berlin, using the Denmark Strait to gain access to the North Atlantic. Several times convoys were sighted but each one had a battleship escort and following orders the German battleships did not engage. They stayed at sea for almost two months and sank eight solitary merchant ships totalling 49,300 tons before reaching Brest on 22 March 1941. Once the pair reached Brest, the battleships were under constant air attack, *Gneisenau* being torpedoed by a Beaufort bomber while lying in the dock basin. Admiral Lütjens was transferred to be flag commander of the ill-fated *Bismarck* and *Prinz Eugen* sortie. After June 1 when *Prinz Eugen* arrived at Brest it was decided that the concentration of heavy units offered too much of a target. *Gneisenau* was still being repaired after the torpedo hit so *Scharnhorst* was moved to the port of La Pallice. The British had observed the move and sent twelve Halifax heavy bombers after the ship on 24 July. Although two of the bombers were brought down they scored five hits on the starboard side. Three of the bombs passed all the way through the hull while two of the bombs exploded on the second armoured deck. With no dockyard at La Pallice, *Scharnhorst* went back to Brest for repairs. During these repairs new weapons were added in the form of torpedo tubes and Vierling quad 20mm mounts. After the loss of *Bismarck* and the subsequent British sweep of German replenishment ships, further North Atlantic sorties were abandoned. Instead, plans were prepared to get the ships back to the north with eventual deployment in Norway in mind.

THE CHANNEL DASH

Operation Cerberus, better known as the Channel Dash, was the last time the *Scharnhorst* and *Gneisenau* worked together as a unit, as after their return to Germany their careers separated. There were two possible routes back to Germany. One was a wide sweep into the Atlantic in a reverse of the Atlantic break-out, which would probably mean an encounter with Royal Navy heavy ships. Another negative factor was the fact that the replenishment ships that had provisioned and refuelled the break-outs of 1940 and 1941 had mostly been picked off. Going through the English Channel placed the ships under concentrated air attack but no heavy ships would be encountered. It was far shorter and the Luftwaffe could provide fighter cover from start to finish, so it was deemed that steaming through the English Channel at high speed presented less risk. Timing was critical, as the longer it took the British to notice the movement of the ships, the less time the ships would be in jeopardy.

Below: The torpedo hit from HMS/M *Clyde* on 20 April 1940 blew a hole right through both sides of *Gneisenau*'s bow and it took a month of temporary repairs at Trondheim before the ship could sail for Kiel. She is shown here in the Howaldtswerke dry dock where repairs lasted until late October 1940. Note the shape of the bulbous forefoot.

Left: *Gneisenau* in a Norwegian fjord near Bergen waiting for orders to break out into the Atlantic. The first attempt in December 1940 was called off when the ship sustained storm damage, but she eventually sailed with her sister ship on Operation Berlin on 28 January 1941. The after end of the hull is in darker grey than the rest of the ship, the turret tops are a dark colour (probably red) and the mainmast is painted black.

Left: During Operation Berlin, the longest sortie by the sisters, there were a number of rendezvous with U-boats in mid-Atlantic. This photograph of *Gneisenau* was taken from *U124* on 6 March 1941.

Left: Showing the effects of weeks at sea, a rather worn-looking *Scharnhorst* arrives off Brest on 22 March 1941. As a raid on commerce the operation was a success, but in choosing to make their destination Brest, which was well within the range of RAF bombers, the ships were exposed to frequent and heavy air raids.

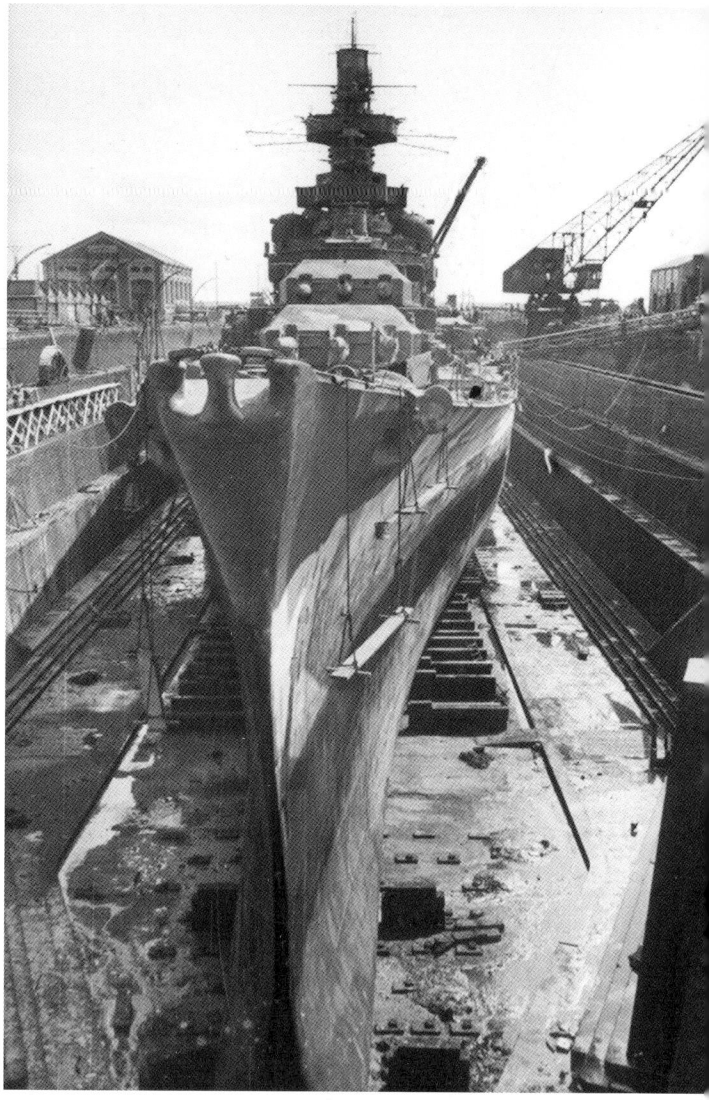

Above: An aerial reconnaissance photograph taken in December 1941 shows the ships in parallel dry docks, *Gneisenau* at the top. Despite the elaborate efforts at disguise, it was never possible to hide the ships completely nor to prevent identification.

Right: *Gneisenau* in a Brest dry dock shortly after her arrival from Operation Berlin. The first RAF raid near-missed the ship, but the threat was enough to persuade the Germans to move the ship out into the yard basin. The RAF seized the opportunity and in a daring raid on 6 April the *Gneisenau* was torpedoed by a Beaufort of 22 Sqd and had to be returned to the dry dock once again.

Below: At Brest the insecurity was compounded by fears that the Resistance were supplying the British with intelligence. To make this more difficult even the recognition codes on the ships' aircraft were painted out (note the different-coloured patch just aft of the cowling where the squadron badge used to be). In the background the *Scharnhorst* is swathed in camouflage netting.

Preparations included keeping the path clear of mines, so the minesweeping command went into high gear. The command had five flotillas of minesweepers and the British noticed the increased activity, although they did not know the purpose, but the British stepped up their own minelaying in response. Escorting the heavy ships were six destroyers and nine torpedo boats (destroyer escorts). At 20:30 on 11 February 1942 the ships raised steam in order to make the maximum use of the hours of darkness. Actual departure was delayed by an air raid but the ships, under the command of Vice Admiral Ciliax, left Brest at 22:45 and remained unnoticed by the British, largely through luck. The ships turned north and rang up 30kts.

When dawn arrived they were well up the Channel. They picked up their fighter escort and kept going north with no appearance by the British. Air power was not the only threat to the German battleships: there were mines. At dawn a flotilla of minesweepers went to work on one of the known minefields through which the battleships would need to pass, exhibiting the thorough advance planning and synchronization for the operation. At 11:20 the battleships reached the known field and slowed to 10kts in order to allow the minesweepers to advance ahead of the heavy ships. They crossed the field undamaged and again increased speed. By noon they were in the Dover Strait and still had not been attacked.

Eventually the British woke up to what was happening. A fighter reconnaissance flight reported a large number of ships and the British mobilized every asset available to go after the German force. It was already past noon and the German ships were now in the North Sea and increasing distance from the British airfields. In all, the British launched over 600 sorties from which 25 bombers and 13 fighters were lost. The aircraft were parcelled into 110 missions and carried a total of 500 tons of bombs. Around 13:45 the first air attack came in with six Swordfish carrying torpedoes; all were lost and no hits were scored. Shortly after entering the North Sea the German escort picked up an additional five destroyers and a horde of smaller escorting craft, again showing the advanced preparation of the operation to maximize the anti-aircraft guns around the battleships at the point of greatest danger. Fighter cover was also at its greatest strength as fighter squadrons were rotated over the battle group.

So far everything had gone perfectly for the Germans but now as the British air attacks were beginning the wind increased velocity and the minesweepers tasked to sweep for mines off the Dutch coast had to break off their operations and return to port and this would have a more serious consequence than the torpedo attacks. By 15:00 the German ships had cleared an area of sandbanks that reduced their speed and they were still undamaged. All seemed well, as the ships had passed the point of greatest danger, and the crews were starting to relax with thoughts that they were home free. At 15:31 that illusion was shattered by a heavy explosion on the starboard bow of *Scharnhorst*. The ship had struck an unswept mine. Engines were stopped as a quick assessment was made and damage control parties went to work. *Gneisenau* took the lead and four torpedo boats (actually the size of small destroyers or destroyer escorts) were tasked to assist *Scharnhorst*. The destroyer *Z29* was ordered alongside *Scharnhorst* to remove Admiral Ciliax and his staff so that he could catch up with the rest of his force, which was quickly leaving *Scharnhorst* behind.

At 16:43, while Ciliax aboard *Z29* was coming up astern of *Prinz Eugen*, the force was attacked by five British destroyers that launched torpedoes at a range of 3000–4,000yds. No hits were scored and one of the attacking destroyers, HMS *Worcester*,

Left: Flying the flag of Vice Admiral Ciliax, *Scharnhorst* leads the German squadron into the English Channel in the famous Operation Cerberus. Some of the strong escort of destroyers and torpedo boats can be seen in the photo taken from *Gneisenau*.

Right: Operation Cerberus: *Gneisenau* preceded by a torpedo boat, one of the 2nd and 3rd Flotillas that joined the escort off Cherbourg. Judging by the wake, the battleship is executing a sharp zig-zag in her course.

was badly damaged. Meanwhile the initial repairs on *Scharnhorst* had been completed and the battleship and her four escorts increased speed to 27kts while a cloud of German fighters provide air cover. Shortly before 17:00 *Scharnhorst* came under heavy air attack. No hits were made and a further hammer and anvil torpedo attack by two Beauforts coming from the stern was evaded. The lead torpedo boat *T13* was badly damaged by near misses and came to a stop and had to be left behind. Around 18:00 the *Scharnhorst* and her three torpedo boats started to encounter damaged German escorts trying to reach safety in a Dutch port. Of these a destroyer and torpedo boat were left behind and then two more German destroyers were seen directly ahead. Ironically, one was the *Z29*, which was letting off steam. *Scharnhorst* had caught up with Admiral Ciliax, who had left the ship two and a half hours earlier. A whaler was being lowered to transfer the Admiral and his staff from *Z29* to the *Hermann Schoemann*. No signals came from *Z29* so *Scharnhorst* rushed past the pair of destroyers.

It was now dark and in the heavy seas the torpedo boats could not keep up with the fast pace of *Scharnhorst*. Only *T16* could be seen, gamely trying to follow in the wake of the battleship. At 19:00 in almost total darkness another air attack came in, but only two aircraft released their bombs, which missed by a considerable margin. Fifteen minutes later the *Hermann Schoemann*, flying Admiral Ciliax's flag, came up from astern and took position ahead of *Scharnhorst*. At 22:15 *Scharnhorst* was in a squall when the first lightship marking a safe coastal channel was spotted. Then another huge explosion knocked out all the battleship's power: *Scharnhorst* had hit another mine. With no power and drifting in the heavy seas, *Scharnhorst* was caught off a lee shore with the wind pushing her towards eventual grounding. *T16* caught up and then *Hermann Schoemann* appeared off the bow. Ciliax signalled that tugs had been dispatched to aid *Scharnhorst*, but by 23:00 *Scharnhorst* was only two miles from shoal

Right: Operation Cerberus: *Scharnhorst*, underway again after striking the first mine, is down by the head but making good speed. The photo was taken from the destroyer *Z29* to which Ciliax had transferred his flag.

water and with no tugs in sight Ciliax decided to tow *Scharnhorst* with *Schoemann*. Two attempts to pass a towline failed but as a third attempt was about to be made one of *Scharnhorst*'s engines became available. Ten knots seemed to be salvation, as a tow from *Schoemann* would barely have made headway. Another squall struck and the ships lost sight of each other. *Schoemann*, using her searchlight, could not find anything, but eventually ran into a heavy oil slick. There was fear that the *Scharnhorst* had gone down but it was noticed that there was a trail of oil leading into the safe channel. The destroyer picked up speed and soon came up on the stern of *Scharnhorst*, which was making noticeable progress towards safety.

At 03:50 the next morning Ciliax heard about the fate of *Gneisenau*. She too had struck a mine in the same area as *Scharnhorst* but had not lost power so was able to reach the anchorage at Helgoland. *Gneisenau* and *Prinz Eugen* had been separated in the heavy squalls encountered by *Scharnhorst*. The cruiser's navigational equipment had failed and she had to take soundings by hand lead-lines all through the night to get to port. Ciliax reboarded the *Scharnhorst* at 08:00 on 13 February 1942, but it was not until noon that all ships had reported in.

THE END OF THE *GNEISENAU*

Scharnhorst and *Gneisenau* had operated together since the start of the war but now fate and negligence took a hand to permanently separate them. *Gneisenau* was the least damaged of the two. She had taken a mine in the stern but her power plant was little affected. On the morning of 13 February she was ordered to the Kiel dockyard for an inspection of the damage, with repairs to be done in the Baltic out of the range of the RAF. It is another irony that the Germans, so thorough in their planning of the escape from Brest, should be slipshod in their planning afterwards. Standing Operating Procedures (SOP) for a ship entering dry dock was to unload its ammunition first. With *Gneisenau* the German navy ignored its own SOP and the battleship went to dock with a full ammunition load. The RAF paid a call on Kiel during the night of the 26/27 February and one bomb found *Gneisenau*. It penetrated the forecastle, went through two decks and exploded on the interior side of the armour belt. The ventilation system sucked hot gases into the ammunition handling room below 'A' turret, which in turn ignited the propellant charges in the handling room below the turret and above the magazine. As these charges immediately exploded, the heavy turret was lifted up and thrown off the roller bearing guides. Only the prompt flooding of the magazines below averted total destruction of the ship. Nevertheless, *Gneisenau* had suffered grievous damage and the bow would need to be completely rebuilt while 'A' turret was also beyond repair. The navy had always recognized that the Achilles' heel of the design was their weak main armament. They already had plans in hand to replace the triple 11in gun turrets with twin 15in (380mm) gun turrets and they had spare 15in turrets available from the cancelling of the 'O' class battlecruisers.

This was an opportunity to carry out the plan and *Gneisenau* was sent to Gotenhafen in the Baltic to have the new guns fitted. She arrived there under her own power on 4 April 1942, was decommissioned on 1 July, and work began with the removal of the 11in gun turrets. The Battle of the Barents Sea on 31 December 1942 changed the plans. A superior German force, including *Admiral Hipper* and *Lützow*, had been repulsed and retreated before British destroyers escorting a Russian convoy. Hitler was furious and ordered an end of all operations by the heavy surface ships and that their crews be transferred to other duties. Work on *Gneisenau* stopped; her two good 11in gun turrets were sent to Norway as coastal artillery, and the 15in turrets that she was to receive went to Denmark and Holland for the same purpose. *Gneisenau* remained at Gotenhafen for the rest of the war serving as a storeship and sometime air-raid shelter. In 1945, as the Red Army drew near, the hulk was towed out and sunk as a blockship on 27 March. After the war Poland dismantled the superstructure and on 12 September 1951 raised the hull, which was scrapped.

Above: A British aerial reconnaissance photo of *Gneisenau* at Gotenhafen in August 1942 when work was still being carried out on the ship. All the 11in turrets have been removed and the new extended bow is under construction. Work was to be halted by Hitler's order at the end of the year, and the twin 15in turrets were never shipped.

SCHARNHORST IN NORWAY

It would take five months to repair the damage of the two mine strikes on *Scharnhorst*. As she went out for trials in July, the state of the engines remained unsatisfactory. The balance of 1942 was used for adjusting the machinery and various trials in the Baltic. Even on 22 December 1942

Right: *Scharnhorst* lying in a Norwegian fjord in 1943. The ends of the hull are painted in a lighter grey but the silhouette is confused by the splinter camouflage of the destroyer *Z33* which is lying alongside.

when she was pronounced ready for operations, she could only steam on two engines with a maximum speed of 25kts. Nevertheless, it was decided that *Scharnhorst* would go to Norway. On 7 January 1943 *Scharnhorst*, *Prinz Eugen* and five destroyers left for the far north. They soon returned to the Baltic as their trip was aborted when they were discovered in transit by the RAF. Later that month, while attempting to avoid ramming a U-Boat, the battleship ran aground off Gydnia and had to go into dock again. The Battle of the Barents Sea also cast a shadow over the future of the *Scharnhorst*, as it had with her sister. Dönitz, the U-Boat commander, was appointed the head of the navy and at first it was decided only to retain the large surface ships then in Norway. However, Dönitz reconsidered his new command and received an exception from Hitler in the case of *Scharnhorst*. *Scharnhorst* left Gydnia for Operation Paderborn on 8 March 1943 with four destroyers and picked up more on the voyage. Heavy storms were encountered off southern Norway and the smaller escorts began to drop astern and seek shelter in a Norwegian port. By the time she reached Narvik on 14 March only two of the escorting destroyers were still with *Scharnhorst*. However, the bad weather had also been a friend as it prevented detection and air attack.

During the summer *Scharnhorst*, *Tirpitz* and *Lützow* carried out some exercises but infrequently due to a shortage of fuel oil. From 6 to 9 September *Scharnhorst*, *Tirpitz* and nine destroyers did undertake Operation Eisberg, which amounted to shooting up some shore installations on Spitsbergen. When the British X-Craft midget submarines attacked on 21 September 1943, *Scharnhorst* had also been targeted. Three, *X5* to *X7*, were to attack *Tirpitz*, *X8 Lützow* and *X9* and *X10 Scharnhorst*. *X8* and *X9* were lost in transit and *X10* had to abort the attack, although *Scharnhorst* had left the anchorage that morning for anti-aircraft exercises in any case. After the attack *Tirpitz* was immobilized and *Lützow* returned to Germany, leaving the *Scharnhorst* the sole remaining heavy combatant active in the north.

On 22 December 1943 a nineteen-ship convoy bound for Murmansk was first spotted and at 17:00 on the 25th the *Scharnhorst* under the command of Rear Admiral Bey set out on her final sortie along with five destroyers in Operation Ostfront. The British quickly had word of the sailing and rushed in the heavy covering force of HMS *Duke of York* (ten 14in guns), the heavy cruiser *Norfolk*, three light cruisers and eight destroyers. The weather was bad with heavy seas, rain and snow forecast. On her approach to the convoy, *Scharnhorst* was sighted twice and one 8in shell from *Norfolk* knocked out the battleship's primary radar on her foretop. At 13:00 on 26 December *Scharnhorst* turned to the south to steam back to Norway. Because of the heavy seas Bey had already released his destroyers to return to Norway and he was sailing blind against a force of unknown strength, but clearly one that did not mind engaging *Scharnhorst*. He had had enough but still did not realize that he was being stalked by a far more formidable opponent than cruisers. *Duke of York* picked up *Scharnhorst* on radar at 14:17 at 25 miles range, but this quickly closed to 17 miles and at 16:40 *Duke of York* ordered the cruiser *Belfast* to illuminate *Scharnhorst* with star shells. The battleship turned broadside and opened fire with superb accuracy at an almost point blank eight miles.

When the star shells burst over *Scharnhorst*, Bey was still unaware of the British battleship and totally unprepared for the impending maelstrom. The first salvo of 14in shells bracketed *Scharnhorst*, including knocking out 'A' turret. *Scharnhorst* had no radar and turned to the north to escape. She ran into *Norfolk* and *Belfast* so turned to the east and tried to use her higher speed to clear the trap. By 15:42 it appeared that *Scharnhorst* might well elude death to fight again. Range was increasing and only *Duke of York* could still hit her. 'B' turret had also been knocked out but as long as she had her machinery working, *Scharnhorst* could outrun and lose *Duke of York*. However, it was not to be. A shell wrecked a boiler room and reduced speed to 10kts. Although 20kts could be restored, it was not enough. A coordinated destroyer torpedo attack scored four hits by 19:00, and five minutes later *Duke of York* and the light cruiser *Jamaica* reopened fire at six miles. 'C' turret returned fire for eleven minutes until it too was knocked out. At 19:30 *Duke of York* ceased fire and *Jamaica* and two destroyers scored a total of eight more torpedo hits. At 19:45 her magazines exploded and *Scharnhorst* disappeared with only 36 of her crew of 1968 rescued.

Model Products

The *Scharnhorst* and *Gneisenau* have not been as popular with kit producers as the *Bismarck* class despite their more active careers. Nonetheless, there are numerous models of *Scharnhorst* and *Gneisenau* ranging in size from a wargaming 1:2400 scale up to 1:350 scale with the Dragon *Scharnhorst*, which is one of the best 1:350 scale kits ever produced.

REVELL MINISHIPS ***GNEISENAU*** 1:1200 Scale

If you are old enough, you may remember a series of plastic kits of Second World War warship models released in a constant 1:1200 scale by the British firm of Eaglewall in the early 1960s. For many of us, these introduced us to this collectors' scale before we encountered the highly detailed, much more expensive, metal models from Neptun. The last and rarest of these kits was a *Gneisenau*. Later Revell revived the concept and produced the Miniship series. Shown here is the Revell Miniship *Gneisenau*. They also produced a separate *Scharnhorst* but it was no different, as both models actualy depicted the *Scharnhorst* with tripod mainmast aft of the hangar. The model is rudimentary with a catapult on 'C' turret and no outboard legs to the tripod. There are raised over-sized portholes on the hull but at least they have the angled anchor channels for the bow anchors. Deck planking is far too wide whereas at this scale one should not notice the planking at all. All anti-aircraft mounts are moulded on the deck, as are also the torpedo tubes and most of the ship's boats. The sole advantage of this model is that it is significantly less expensive than the metal models of this ship in 1:1250 scale by Neptun and 1:1200 scale by Superior/Alnavco.

Sample: **Author's Collection**

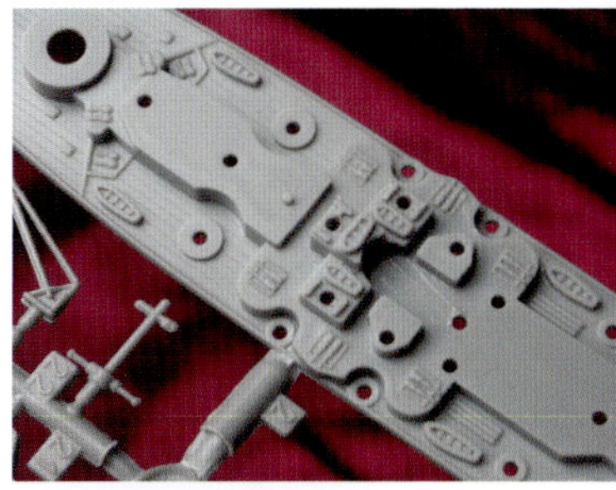

Mid-size Scales 1:550 to 1:720 Scale

This scale range is by far the most popular, and for the *Scharnhorst* class there is a wide range of kits available scattered among different scales. The *Scharnhorst* class has been left behind when it comes to mid-size models and, oddly, the best models in mid-size for *Scharnhorst* is in 1:600 scale by Airfix and in 1:700 scale by Tamiya, each over thirty years old. One other 1:700 model not included in these reviews but seen in the gallery, is a resin 1:700 scale model of *Gneisenau* as she would have appeared with twin 15in gun turrets after the planned reconstruction of 1943. This model is produced by H-P Models.

REVELL ***SCHARNHORST*** 1:570 Scale

The Revell AG *Scharnhorst* has been around for more than a quarter of a century. Its large size in 1:570 scale certainly allows room for detailing. The model depicts the ship after the modifications that altered her appearance from that of *Gneisenau*, with the enlarged aircraft hangar and tripod mainmast mounted aft of the hangar. It is a full hull kit, but Revell does have a line on the hull pieces

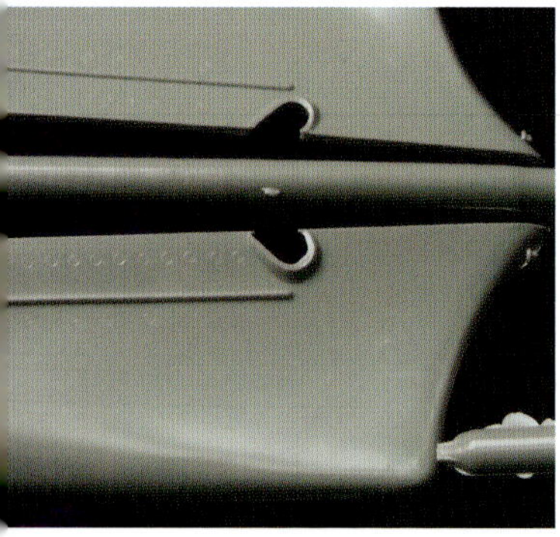

denoting the waterline if you want to remove the lower hull. For a large model hull its detail is somewhat sparse. The overall lines are good with correct openings for the anchor channels but there are horizontal strakes at bow and stern that are not only over-size but were never on the ships. The anchor on the port quarter is simplified with the anchor moulded onto the hull, rather than a separate anchor within an anchor well. The one-piece main deck is also lacking in detail, given the size of the model. Revell went to the effort of providing open anchor recesses on the hull sides but they will just be black holes, as there are no channel parts leading to the forecastle. The bow anchors are moulded flat on the deck. Deck planking is delineated with raised lines, rather than incised plank separation lines. Ship's boats are moulded to the main deck instead of being separate parts and there are no support brackets for the breakwaters. The shelter deck reflects wooden planking instead of a metal anti-slip pattern deck. Many small parts miss detail as well. The 4.1in gun mounts could have been far better: as it is, they are just square blocks with moulded-on gun barrels. The single 20mm guns are thicker and the 37mm mounts simpler than those provided in the smaller 1:600 scale Airfix kit. The solid aircraft cradles are also sad when compared to their equivalents found with the *Scharnhorst* from Airfix. This Revell kit is an enigma. Although produced at least fifteen years after the Airfix kit and in a somewhat larger scale, the Revell version is clearly inferior to its Airfix competitor in detail, although it is in the better known Atlantik bow configuration instead of the as-built version of the Airfix kit.

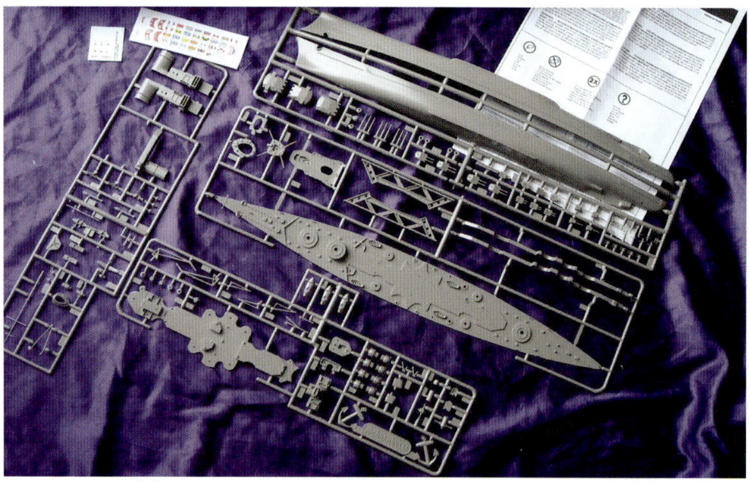

■ Sample: **Author's Collection**

REVELL *GNEISENAU* 1:570 Scale

All of the comments about Revell's *Scharnhorst* apply to their *Gneisenau* as well. Most of the parts are shared in common between both models, except that *Gneisenau* reflects the earlier fit of both sisters with the tripod mast behind the funnel but ahead of a smaller hangar. The problem is that the Revell *Gneisenau* does not have the smaller hangar – the *Gneisenau* hangar parts are identical to the lengthened hangar received only by *Scharnhorst* and not by *Gneisenau*, at least not until autumn 1941. The parts that are different are the shorter tripod mast and a catapult and platform for attaching to the crown of 'C' turret. The portholes on the

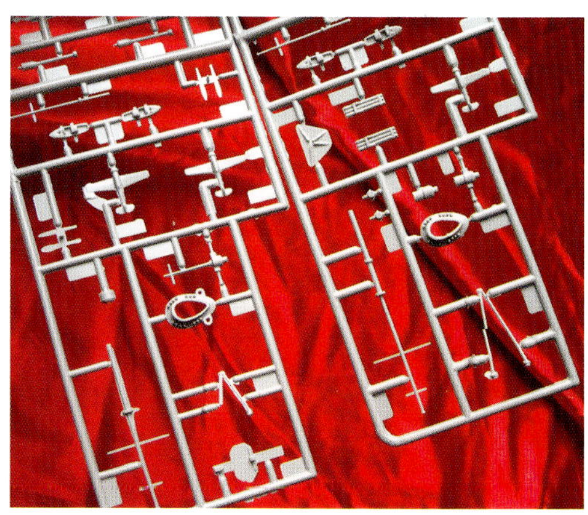

hull and superstructure are far too large and there is no indication of the unique German square superstructure windows. The best bet to improve either the Revell *Gneisenau* or *Scharnhorst*, short of scratch-building, is the White Ensign Models 1:600 scale photo-etch Kriegsmarine set. It will not fix some of the more glaring errors, such as the *Gneisenau* hangar/catapult or lack of anchor channels connecting the forecastle to the hull sides, but at least it will do something to overcome the austere appearance of the kits as supplied.

Sample: **Author's Collection**

AIRFIX *SCHARNHORST* — 1:600 Scale

Early Airfix output concentrated on British warship subjects, followed by German ships sunk by the Royal Navy. Among these the *Scharnhorst* stands alone in presenting the initial, as built, fit with straight stem and flat topped funnel, which shows the original family resemblance to the panzerschiffe, before the addition of the Atlantik bow and raked funnel cap. The full hull kit features open anchor hawse holes of the traditional variety before the narrower Atlantik bow required relocating the anchors in chutes slanting down from the deck edge. The bow also has raised shield outlines for the ship's coat of arms. The Airfix *Scharnhorst* is one of the later kits produced, so that even though it was produced over forty years ago, there is still significant hull detail such as open chocks, raised porthole fittings and locator holes for boat booms. The fore and aft main decks feature fine plank detail but with raised, rather than incised, plank separation lines and no butt end detail. The shelter deck does not have the anti-slip deck pattern. Other deck detail includes bollard fittings, moulded-on anchor chain, anchor locker fittings and some hull access hatch coamings. Also unique to the as-built fit, the Airfix kit has a catapult with aircraft platform on the crown of 'C' turret and includes miniature Heinkel He 60 biplane floatplanes that equipped the large warships of the German navy before the Second World

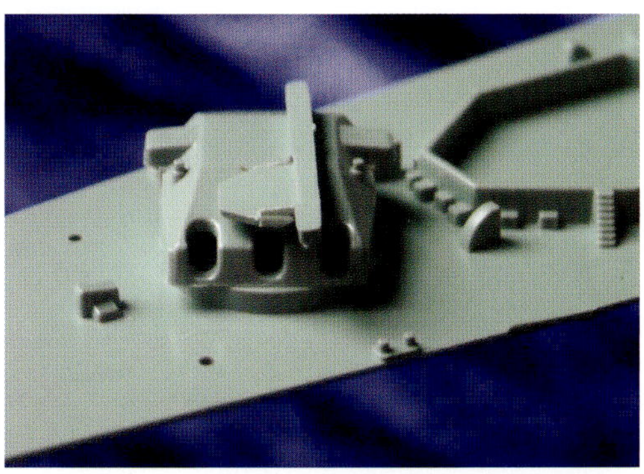

War, rather than their replacements, the Arado Ar 196. Now, these little aircraft are not well detailed, but they are there. The kit has some finely done smaller parts in spite of its age. The single 20mm AA guns, twin 37mm mounts, separate crane and crane arms, anchor windlasses, separate navigation wings (allowing the modeller to portray them as stowed against the superstructure or deployed out from the superstructure) are all remarkably nice for a kit of this vintage. My only real complaint concerns the moulding dimples on the single 5.9in turret crowns and ship's boats, but there is an easy remedy in the form of small dabs of putty and sanding. This Airfix kit is currently not in production but I strongly recommend acquiring a copy the next time that Airfix releases their *Scharnhorst*.

Sample: **Courtesy of Felix Bustelo**

TAMIYA *SCHARNHORST* — 1:700 Scale

The Tamiya *Scharnhorst* has been the workhorse of model kits of this class since its release in 1975. It is now long in the tooth and clearly needs a retooling as was received by *Bismarck* a few years ago by multiple manufacturers. Still, the Tamiya *Scharnhorst* is a good kit and can be readily enhanced by numerous photo-etched and other media after-market parts from a wide variety of producers. In looking at the one-piece waterline hull and deck piece, I was struck by the clean lines of the mould: thirty-seven years after its release and this is still a good kit. Unlike other kits in which

you have to drill the portholes, the Tamiya *Scharnhorst* has them drilled already. Also, unlike other models, which have separate decks and hull parts, the Tamiya kit has most of the main deck moulded as part of the hull. I like this approach as there are fewer seams that may require filling. There is very fine plank decking detail, lacking butt-ends but still holds up well. If you wish, it can be replaced with actual wooden deck (see the accessories section below). The separate forecastle has a couple of problems. For one, the breakwater lacks support brackets and, for another, Tamiya chose to scribe the deck swastika emblem into the plastic. It may make it easier to hand paint but it was a retrograde option, more in line with plastic kits from the 1950s. The Tamiya *Scharnhorst* has a quad 20mm gun position on the crown of 'B' turret. *Scharnhorst* carried this mount during Operation Cerberus but it was removed after the Channel Dash, and the mount did not have a solid splinter shield, so it would be better to remove the gun tub, sand, and replace with railing. Unfortunately the funnel is not open, to give a three-dimensional appearance. It would be better to open the funnel and use a photo-etch grating. In spite of these opportunities for improvement, the Tamiya *Scharnhorst* has a lot of commendable qualities. The bridge superstructure is rather nice with pre-drilled portholes. It has the best 4.1in gun mounts of any of the mid-size kits and the small plastic parts are appropriately thin. Until Tamiya or another company releases a retooled kit, the Tamiya 1:700 scale *Scharnhorst* remains the reigning king of the mid-size kits.

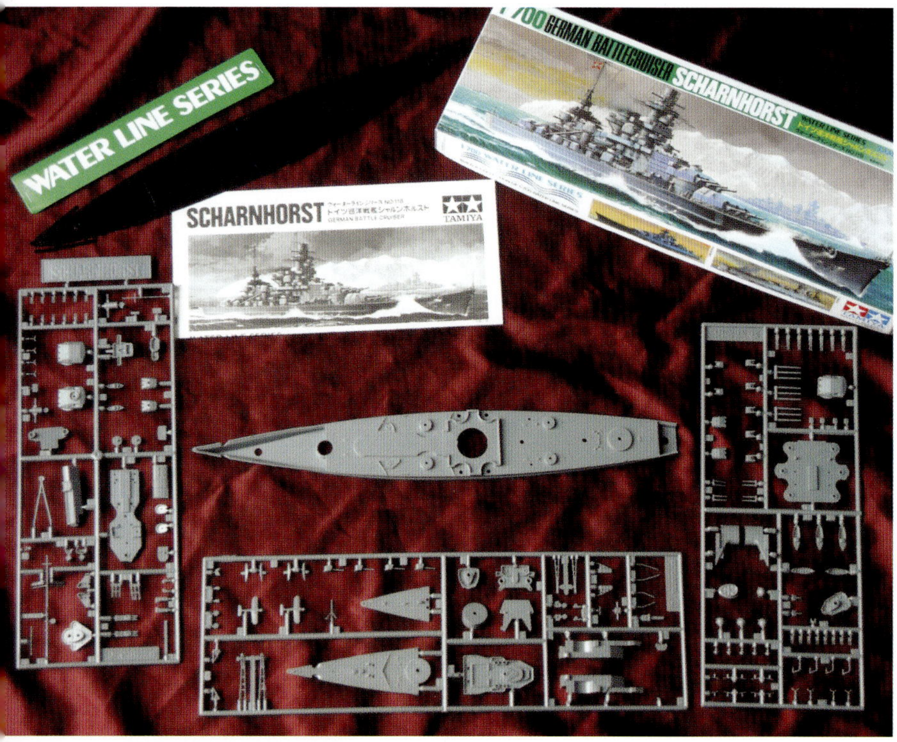

■ Sample: **Author's Collection**

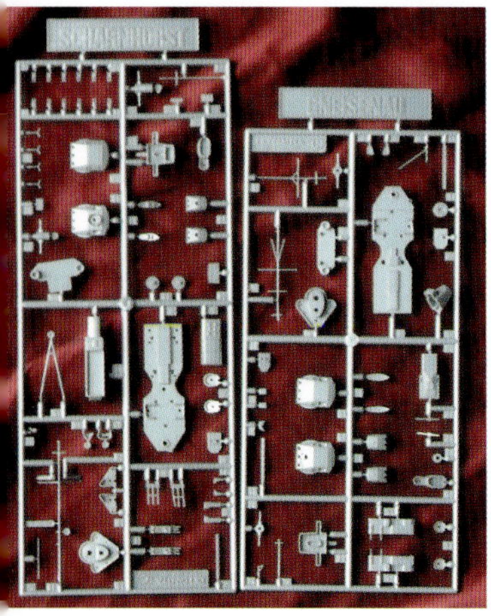

TAMIYA *GNEISENAU* 1:700 Scale

Tamiya also produces a companion *Gneisenau* to the *Scharnhorst*. While the *Scharnhorst* kit appears to be the early 1942 Channel Dash fit, the *Gneisenau* appears in an early war configuration with an aircraft platform and catapult atop 'C' turret. Of course the *Gneisenau* kit has different parts for those areas that distin-

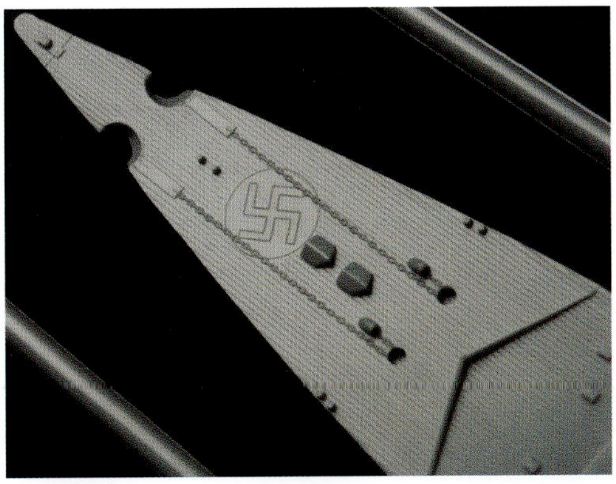

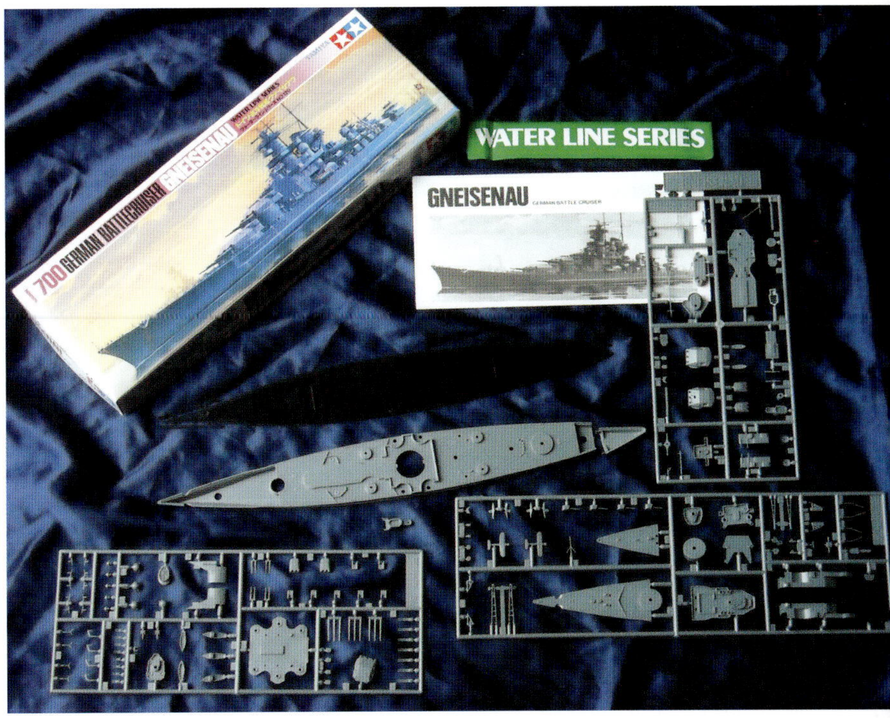

guished *Gneisenau* from her sister, such as smaller hangar and mainmast on the funnel platform. Both kits feature the same odd stern, supplied as a separate piece from the rest of the hull. However, the stern pieces for *Gneisenau* and *Scharnhorst* are identical and the only logic that I can see for the inclusion of a separate stern is to fit in a smaller box than would be required if the stern was part of the one-piece hull. Since the kit was released in 1975, it may be a residual throwback to the box-size models of the 1950s and 1960s when production of the boxes cost more than the production of the plastic parts. Unfortunately for the modeller the decision to include a separate stern necessitates sanding of the seam joining the stern piece with the full hull. This separate stern does have the stern anchor well with separate anchor. Since this is an early war fit, it is the choice to use for the summer 1939 appearance of the ship allowing use of the gorgeous enamelled ship's crests found in the KD Models upgrade set, a review of which is included below.

Sample: **Author's Collection**

LINDBERG *SCHARNHORST* 1:762 Scale

This must be a throwback to half a century ago when some companies produced kits, not in a constant scale, but to fit inside a standard sized box. Who produces kits in 1:762 scale? There is, however, one benefit to this kit – price. It is cheaper than the Tamiya kit and would make a good first kit for a child. It is simple, has no fiddly fine parts and with a one-piece full hull, will float well in any bath tub. For any modeller over the age of ten, the kit has little to recommend it. The detail is woeful! The anchors are moulded on the deck with no anchor chutes. Deck fittings are minimal with moulded-on 37mm and 20mm gun mounts, blocky bollards and flat hatch coamings. The deck planking is over wide but, interestingly, has incised plank separation lines and includes butt-end detail, not found in other, more detailed kits from other companies. The smaller parts are generally rudimentary with some efforts to provide detail – but the detail is questionable, such as the odd ovals on each side of the lower funnel. The 4.1in gun mounts are a total miss, as the kit parts appear as enclosed turrets, rather than open gun mounts. The kit portrays the ship as modified with Atlantik

bow but also includes an aircraft catapult for the crown of 'C' turret, which was removed a few months later. The Arado floatplanes have no floats. Moulding quality is less than perfect as well, as some of the 11in gun barrels and the tripod mainmast were warped.

Sample: **Author's Collection**

Larger Size – 1:400 to 1:350 Scale

There is a small selection of larger scale kits of the *Scharnhorst* class. First and foremost by far is the 1:350 scale Dragon *Scharnhorst*, which is one of the best injection moulded ship kits that you can buy – not only of the *Scharnhorst*, but of any warship. Iron Shipwright also produced a $300+ resin and brass version but production ceased once the much more economical 1:350 scale Dragon *Scharnhorst* appeared, and Heller produced 1:400 scale versions of *Scharnhorst* and *Gneisenau*. At this time it is not known if Dragon will produce a 1:350 scale *Gneisenau*.

HELLER ***GNEISENAU*** 1:400 Scale

This and the Heller *Scharnhorst* have been the models of choice for those modellers wanting a larger model of *Scharnhorst* or *Gneisenau*, until the release of the Dragon *Scharnhorst*. Still, at less than half of the cost of the Dragon version, the Heller kits are at the lower end of the high price spectrum. As a plus, the models can be easily enhanced by a wide variety of after-market parts designed by various manufacturers for these kits. Originally Heller released separate *Scharnhorst* and *Gneisenau* kits but in the latest reincarnation, released by Airfix, it is marketed as *Scharnhorst/ Gneisenau*. The *Gneisenau* kit has a separate sprue for parts specific to that ship and presumably the Heller *Scharnhorst* has a similar sprue with parts unique to that ship. The Airfix release may provide both ship-specific sprues but since I have not seen the contents, I cannot be certain. The Heller *Gneisenau* most closely portrays her appearance from the autumn of 1941 to her arrival in Germany following Operation Cerberus. Throughout her raids in the Atlantic *Gneisenau* had carried a small hangar, which was one of the differences between her and *Scharnhorst*. While at Brest a longer hangar was added to *Gneisenau* late in 1941, with the catapult moved inside the hangar to allow catapulting the aircraft directly from the hangar; consequently there was no catapult on top of the hangar. The *Gneisenau* sprue includes this longer hangar but is in error by having a catapult on top. The kit only has single 20mm guns, so after-market quad 20mm mounts would have to be added, including one on the crown of 'B' turret. Hull lines are clean and a degaussing cable is shown on the hull sides and the scuttles (portholes) are easily opened. Deck detail is unspectacular but competent. A plus is a separate breakwater with support brackets. Superstructure portholes need to be drilled, as unlike the hull scuttles that are drilled most of the way through, the superstructure portholes are flat circles. Navigation wings have solid bulwarks, which could represent canvas-covered railing but a better approach would be to remove the bulwarks and replace with photo-etch railing. The kit comes with plastic rails, which may have been high-tech in 1980 but is obsolete now: don't use it and go with brass photo-etch instead. The Heller *Gneisenau* is still a worthy kit for consideration, as it is large and less than half the price of the 1:350 scale Dragon *Scharnhorst*.

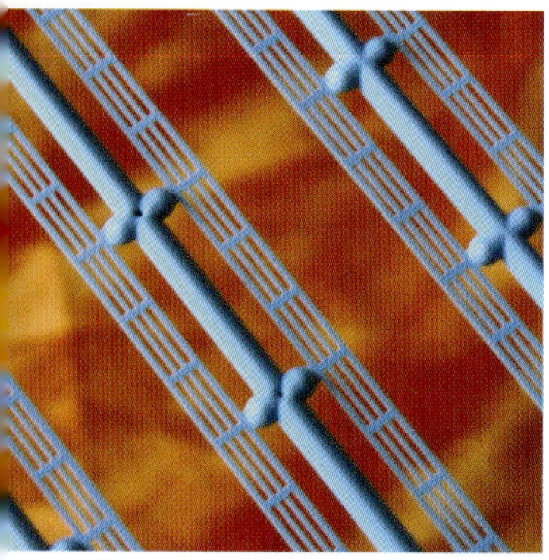

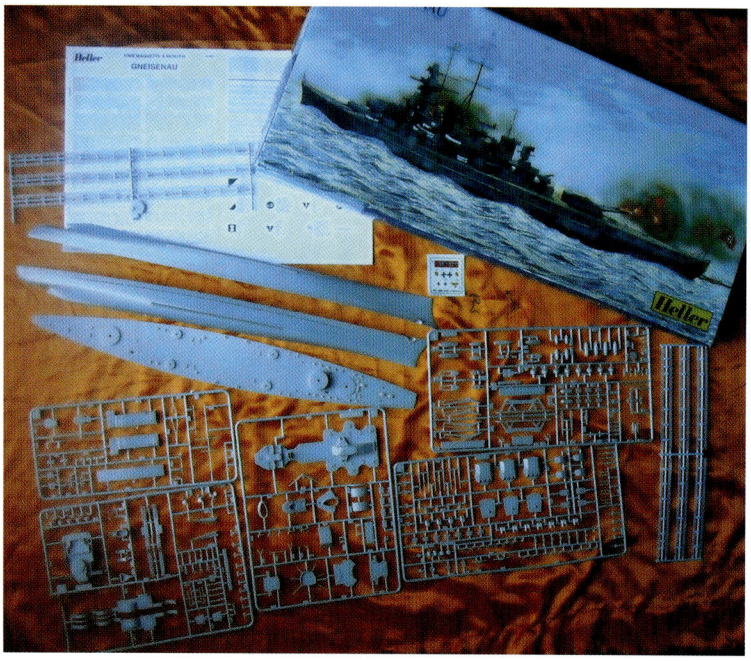

Sample: **Author's Collection**

DRAGON ***SCHARNHORST*** 1:350 Scale

By far the best ship kit that any modeller can acquire in any scale is the 1:350 *Scharnhorst* by Dragon. Nothing but superlatives can describe this model. It comes with a separate lower hull, so the ship can be replicated in full hull or waterline format, and everywhere that you look you are met with endless detail. Anti-slip

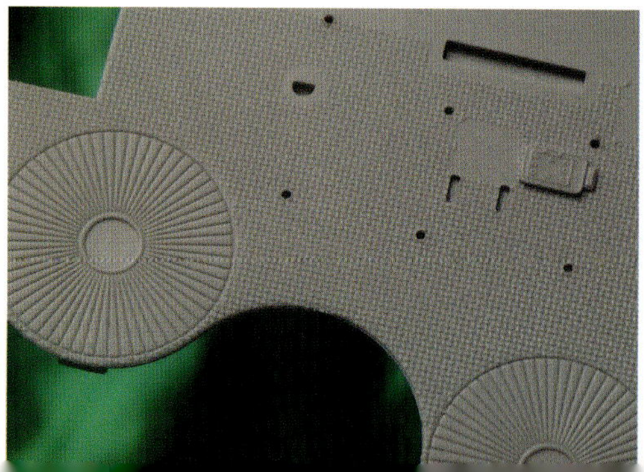

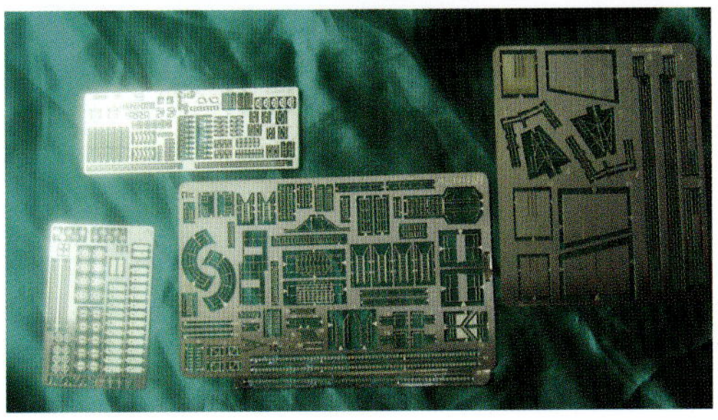

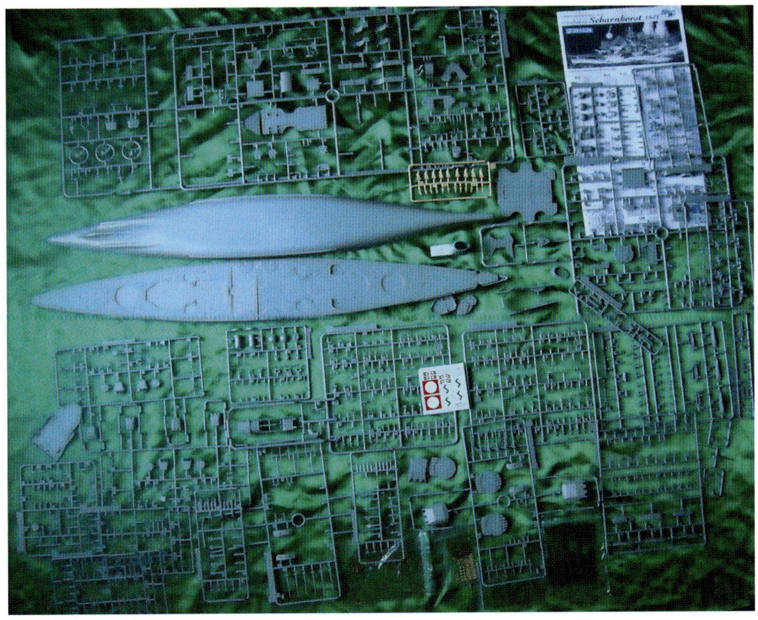

patterns for metal decks are excellent, but I am somewhat surprised that the wooden deck planking does not show butt-end detail. Scuttles have their rigols (eyebrows) and deck access hatches are finely detailed. Hand rungs are present on the hull and superstructure. There are twenty-six sprues of plastic parts and that does not include upper hull, lower hull or stand. The kit even comes with four frets of relief-etched brass. The moulding of the smaller parts is on an equally stellar level. Dragon put some thought into moulding the kit to minimize seams. As cases in point both the bridge superstructure and funnel are one-piece parts requiring no seam sanding. The fit of the kit is listed as 1943 and no Arado Ar 196 floatplane is included. With close to a thousand parts, the Dragon 1:350 scale *Scharnhorst* will not build quickly but its quality is worth your patience.

■ Sample: **Courtesy of Dragon Models**
www.dragonmodelsusa.com

Photo-Etch & Accessories

ARTWOX MODEL WOODEN DECK FOR THE DRAGON *SCHARNHORST* 1:350 Scale

More and more modellers are now using actual wooden decks to replace painted plastic decks. The benefits are obvious with very high quality planking detail with real wood, butt-end detail and an overall superb appearance; the down side is that the wooden decks fit on top of the plastic decks, increasing the height slightly. However, Artwox Model produces a wooden deck for the Dragon *Scharnhorst* kit that is extraordinarily thin, minimizing the increased height. Cut-outs are punched out, allowing the deck fittings on the plastic deck to pass through and appear on top of the wooden deck. Installation of photo-etch railing will allow the bottom rung of the railing to serve as a waterway and hide the wooden deck edge but in spite of its thinness, the deck fittings will have less height above the deck than otherwise. Artwox produces their decks in a sandwich format: wooden deck, adhesive in the middle, and at the bottom a thin plastic protective cover than is removed before actual attachment of the deck. Before you remove this cover, always test fit the wood with the plastic deck to make sure all fitting cut-outs are removed and that there is a good fit without buckling.

■ Sample: **Courtesy of Artwox Model**
http://www.artwox.com/

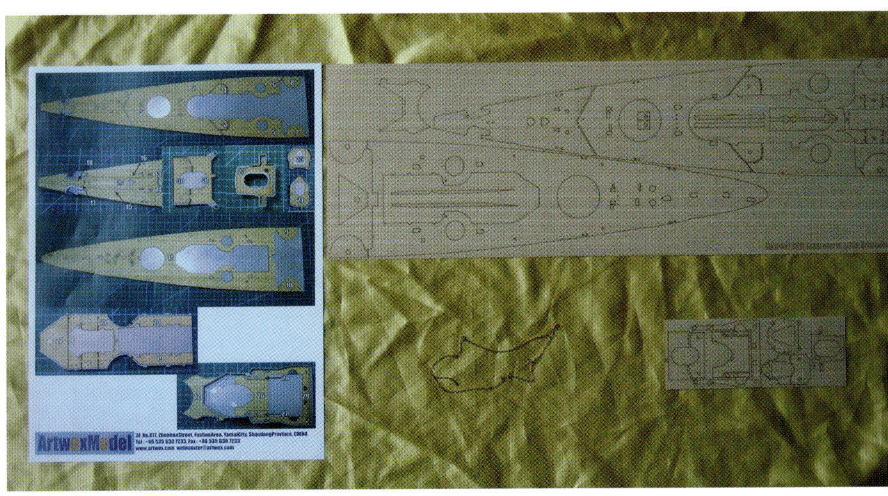

FLYHAWK BRASS REPLACEMENT MAINMASTS

1:700 Scale

Flyhawk produces replacement mainmasts to substitute for what Tamiya include in their 1:700 scale kits. Both sets are highly recommended, especially for the more complex *Scharnhorst* mainmast. The set includes turned brass mast legs and a relief-etched brass fret for platforms, supports and railings. The result is a far finer mast than can be produced from the plastic parts in the kit.

Sample: **Courtesy of Flyhawk Model**
http://www.flyhawkmodel.com

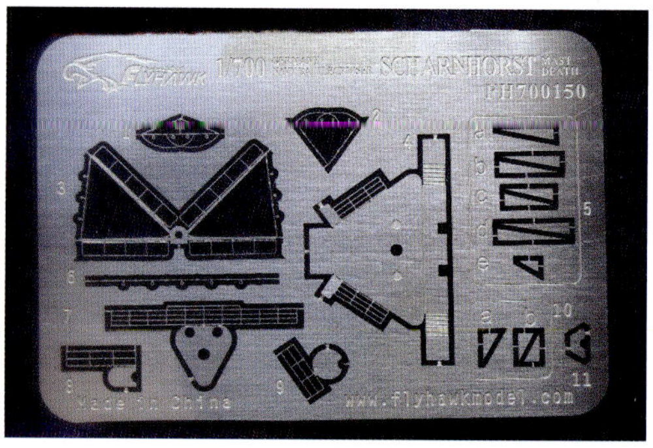

FLYHAWK 37MM/L83 SK C30 ANTI-AIRCRAFT GUNS

1:700 Scale

Flyhawk set FH700130 provides replacement 37mm/L83 SK C30 mounts for the Tamiya 1:700 scale *Scharnhorst* or *Gneisenau*. Twenty-three gun mounts are provided with each mount comprised of five parts. The parts are small and fiddly but the payoff is high with much finer detail for these twin anti-aircraft guns.

Sample: **Courtesy of Flyhawk Model**
http://www.flyhawkmodel.com

FLYHAWK KRIEGSMARINE NAMEPLATES

1:700 Scale

Flyhawk Model set FH700028 presents brass nameplates for the four Second World War German Battleships. With the names in Gothic Font and with relief-etched silhouettes of the ships *Bismarck*, *Tirpitz*, *Scharnhorst* and *Gneisenau*, these plates are very well done. Note how the silhouettes of *Scharnhorst* and *Gneisenau* vary.

Sample: **Courtesy of Flyhawk Model**
http://www.flyhawkmodel.com

GOLD MEDAL MODELS WW2 GERMAN WARSHIP FITTINGS SET

1:400 Scale

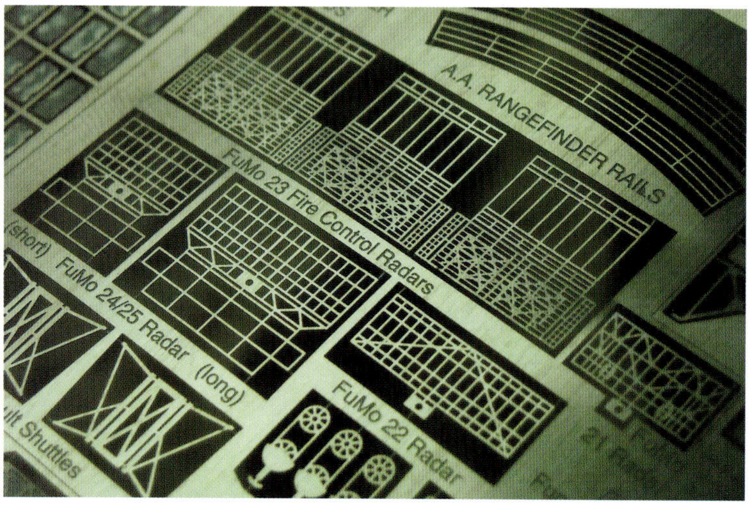

Gold Medal Models has a relief-etched brass photo-etched set designed to equip all of the 1:400 scale models produced by Heller. The parts in this set are exceptionally well done but most of the parts to be used on the Heller *Scharnorst* or *Gneisenau* are generic, rather than ship-specific, but the latter include the foldable bridge navigation wings, funnel grating, radars and mast yardarms. However, there is a wealth of generic parts in this fret that will really pump up the detail on either Heller kit. Included are parts for: railing; doors; cable reels; life buoys; 20mm guns; inclined ladders, vertical ladder; crane pulleys and anchor capstan heads.

Sample: **Courtesy of Felix Bustelo**

KA MODELS DETAIL SET

1:700 Scale

KA Models produces an outstanding super-detail set for the Tamiya *Scharnhorst* and *Gneisenau*. It includes wooden decking and a supplementary set of photo-etch brass. The wooden decks have butt-end detail with pop outs for fittings to pass through. The major brass part is the shelter deck with anti-slip metal pattern, as well as anchor chain to lay atop the wooden deck. The brass ships' crests come ready enamelled with the correct colours.

Sample: **Courtesy of Freetime Hobbies**
www.freetimehobbies.com

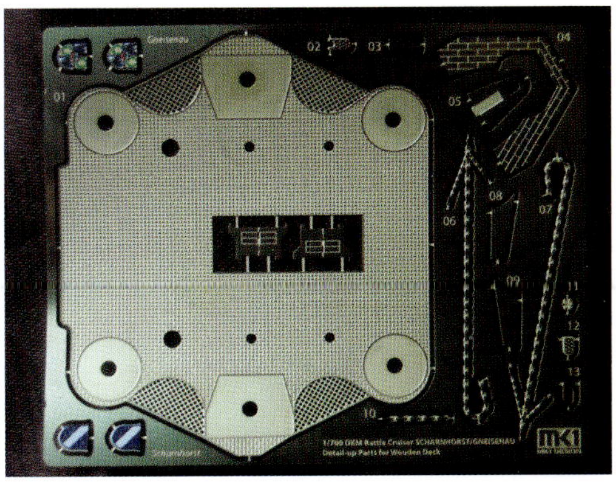

L'ARSENAL ARADO AR 196 FLOATPLANES

1:350 and 1:400 Scales

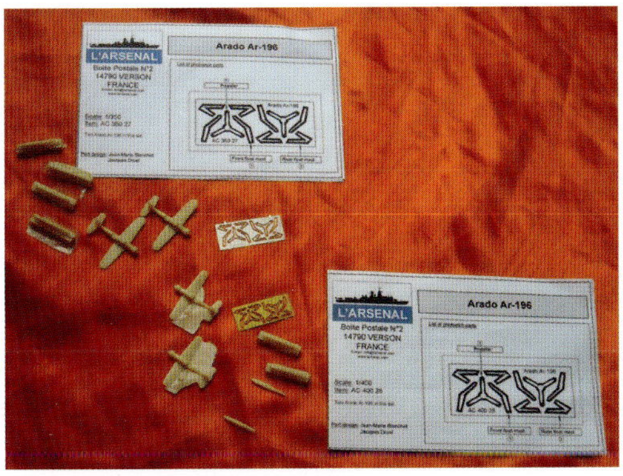

The French firm L'Arsenal produces finely detailed resin and brass models of the standard Arado floatplane with which both *Scharnhorst* and *Gneisenau* were equipped after the start of the war. In each scale, there are two aircraft per bag with resin parts for the aircraft and floats and brass photo-etch parts for the propeller and struts connecting the floats to the fuselage. The combination presents a more detailed and finer appearance than plastic parts.

Sample: **Courtesy of Felix Bustelo**

LION ROAR DETAIL SET FOR THE QUADRUPLE 20MM VIERLING

1:700 Scale

While other navies went for the Swiss-designed Oerlikon, the Kriegsmarine developed their own weapons for light AA defence. In 1940 a new mount was developed to further increase armament. Designed by Mauser and produced by Rheinmetall, the weapon was the very effective quad 20mm formally known as the 2cm Flak 35 im Vierling L/38; this weapon had four C38 guns on a new mount. The German Vierling was an excellent light AA system and was mounted on the *Scharnhorst* or *Gneisenau*. Lion Roar has produced a two-fret set of photo-etched Vierling in 1:700 scale, set LE700013. The larger of the two frets is of brass and contains five different types of parts to assemble the chassis of the mount. The smaller fret is of stainless steel and contains the gun barrels and gunners' seats. A total of sixteen quad 20mm mounts can be assembled from this set, but there are only parts for eleven gun shields.

Sample: **Courtesy of Lion Roar**
www.lionroar.net

LION ROAR DETAIL SET FOR THE CABLE/HOSE REELS

1:700 Scale

As with the ships of other nations, Kriegsmarine warships carried a significant number of cable and hose reels as fittings on the deck or on the side of bulkheads. Although cable reels are found in some resin and brass 1:700 kits, they are very rarely found as detail in plastic kits. If they are present as part of the deck detail, they are more likely to resemble amorphous blobs. Lion Roar has released a brass photo-etch set that will supply you with detailed cable and hose reels for your 1:700 scale German warship kits, including *Scharnhorst* and *Gneisenau*, regardless of whether they are in plastic or resin. Lion Roar photo-etch set LE700008 is devoted just to cable reels of the Kriegsmarine. The brass fret contains reels in three different sizes. There are sixteen large reels, eighteen medium size reels and thirty-six small reels. Although the medium and small reels appear to share the same basic design, the large reels are completely different. All seventy reels in the fret are relief-etched with the reel frame overlapping the reel ends.

Sample: **Courtesy of Lion Roar**
ww.lionroar.net

LION ROAR 37MM/L83 SK C30 ANTI-AIRCRAFT GUNS

1:700 Scale

Lion Roar set LE700014 is an early photo-etch set from this company. Unlike the Flyhawk approach of providing more parts per gun mount, Lion Roar reduces the parts per gun but has the modeller do more folding. I personally prefer the Flyhawk approach.

■ Sample: **Courtesy of Lion Roar**
www.lionroar.net

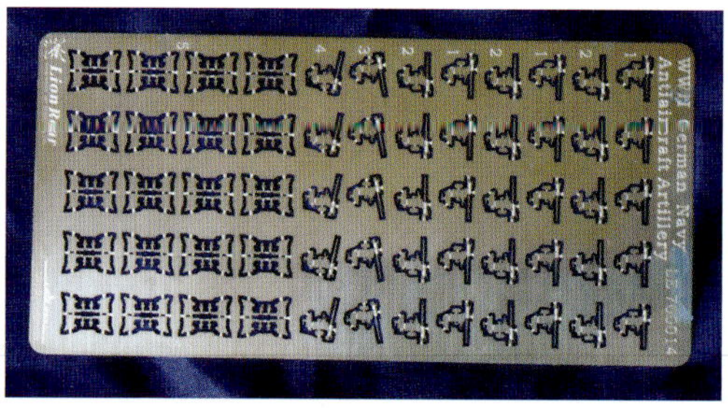

TOM'S MODELWORKS *SCHARNHORST* CLASS

1:400 Scale

Tom's Modelworks has produced a brass photo-etched fret set designed for the *Scharnhorst* and *Gneisenau* kits produced by Heller in 1:400 Scale (Set 4012). There are two frets in the set: the first contains ship-specific items and measures 7½in by 2⅞in; the second contains railing and vertical ladders and measures 6¾in by 2in. Two things that jump out when you open the Tom's package for the *Scharnhorst* and *Gneisenau* are the 01 level safety nets and the various radar arrays. The safety nets fit in the 01 deck semi-circular cut-outs over the main deck 5.9in secondary positions. The mesh of the safety nets, radar screens, various platforms and catwalks brass parts will really supercharge the Heller model.

■ Sample: **Courtesy of Tom's Modelworks**
www.tomsmodelworks.com

TOM'S MODELWORKS INCLINED AND VERTICAL LADDERS

1:400 Scale

Not included in their *Scharnhorst* class set in 1:400 scale were both vertical and inclined ladders. Tom's set 4017 addresses this need with multiple patterns of inclined ladder. They come in five, six, seven or eight treads, with and without weight saving voids in the bottom frames. Also included are four accommodation ladders and four long runs of vertical ladder.

■ Sample: **Courtesy of Tom's Modelworks**
www.tomsmodelworks.com

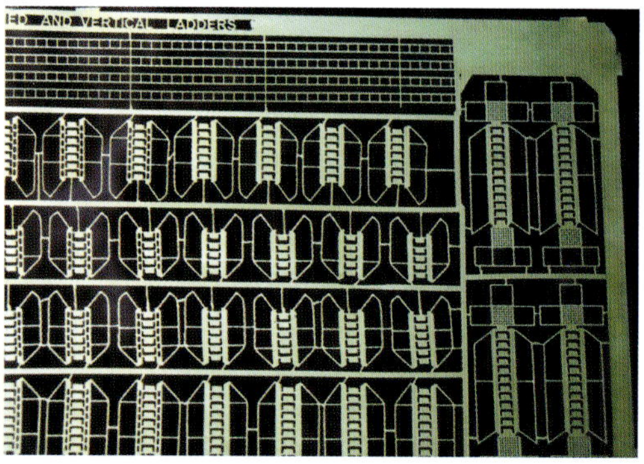

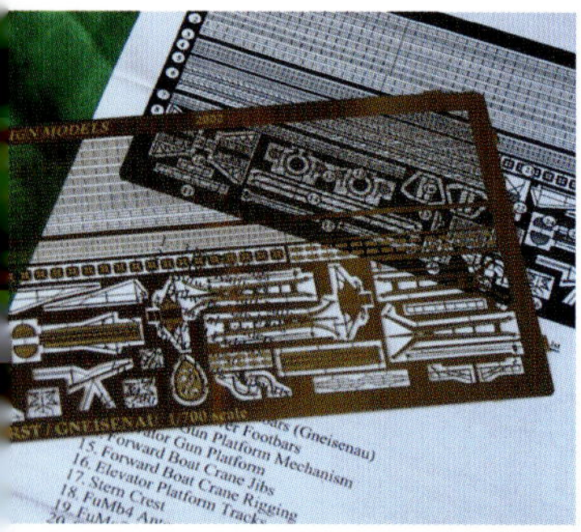

WHITE ENSIGN MODELS *SCHARNHORST* CLASS

1:700 Scale

Released in 2002 the White Ensign Models set for the Tamiya *Scharnhorst* and *Gneisenau* replaces the cranes and catapults. Other ship-specific parts are the radar, folding navigation wings, funnel cap, stern eagle, davits, Arado detail, some doors, and Marx floats. More than 50 per cent of the fret is for railing in different patterns and vertical ladder. No inclined ladder is provided. Overall, this is a somewhat disappointing release from a company known for quality and comprehensiveness.

■ Sample: **Courtesy of White Ensign Models**
www.whiteensignmodels.com

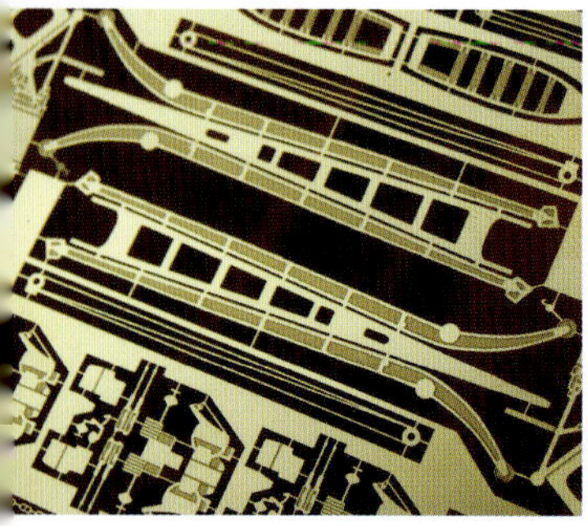

WHITE ENSIGN MODELS *SCHARNHORST* CLASS

1:400 Scale

In contrast to their 1:700 scale equivalent, the White Ensign Models set for the 1:400 scale Heller *Scharnhorst* provides what was missing in the former set. It is odd because the 1:400 scale set was also released in 2002 but provides detail missing in the smaller version, including inclined ladders, the heraldic shields that *Scharnhorst* carried on either side of the bow, windows, boat thwarts, cable reels, 37mm AA guns and tower foretop. The 1:400 scale set also includes everything found in the 1:700 scale fret.

■ Sample: **Courtesy of White Ensign Models**
www.whiteensignmodels.com

WHITE ENSIGN MODELS *SCHARNHORST* CLASS
1:350 Scale

Released in 2010 with the Dragon 1:350 scale *Scharnhorst* in mind, the set resembles a scaled-up version of their 1:400 offering but with even more detail. Crests are included for both *Scharnhorst* and *Gneisenau* and WEM provides single 20mm mounts with separate ammunition clips. A lot more detail is provided for ship's boats as well, with railing, rudders, propellers and navigation wheels. A very nice touch is the bow railing, which is curved to fit the sheer of the deck line.

■ Sample: **Courtesy of White Ensign Models**
www.whiteensignmodels.com

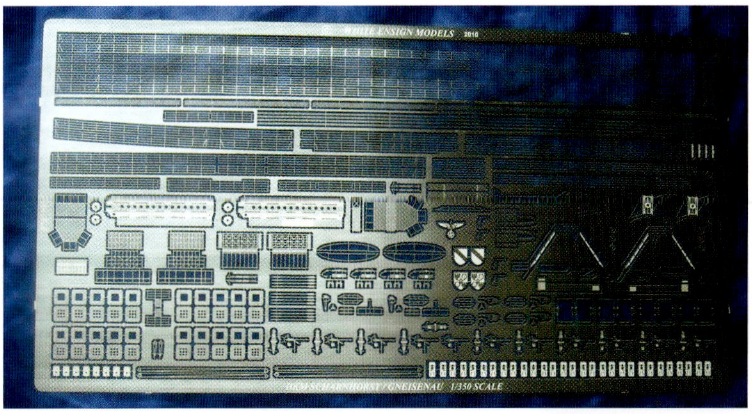

WHITE ENSIGN MODELS DKM ANTI-AIRCRAFT WEAPONS
1:700 and 1:350 Scale

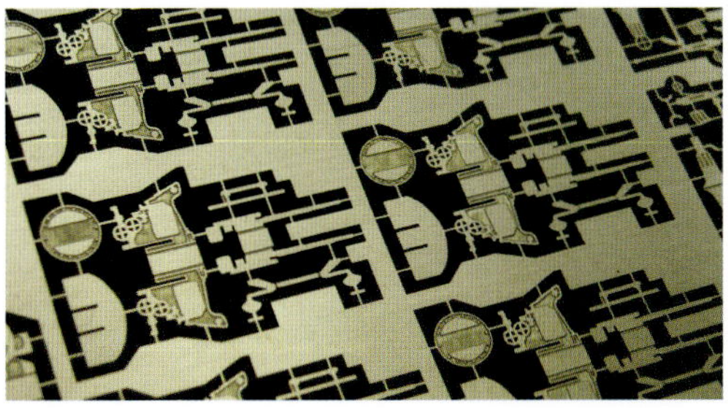

PE 733 is a small fret providing single 20mm guns and the four-barrelled Vierling mounts in 1:700 scale. PE 3566 provides the same mounts found in the 1:700 scale version, except the proper C30 37mm mount are included. Extensive relief-etching is used throughout.

■ Sample: **Courtesy of White Ensign Models**
www.whiteensignmodels.com

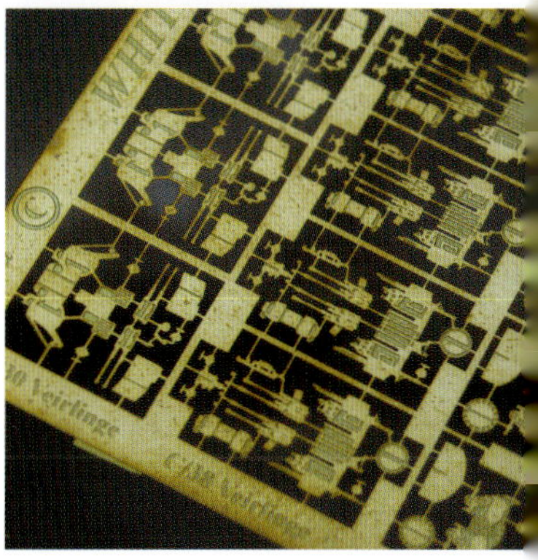

WHITE ENSIGN MODELS 'ULTIMATE KRIEGSMARINE' SET ONE
1:600 Scale

White Ensign Models Set WEM PE 626 is for all of the Airfix 1:600 scale German warship kits, not just for *Scharnhorst*. This large fret still provides a lot of finely detailed parts that will significantly improve the appearance of the Airfix kit. Included are 'C' turret catapult and platform; aircraft cradles; Nazi eagle stern decoration; boat details; various platforms and supports; aircraft cranes; radars; and mast detail.

■ Sample: **Courtesy of Felix Bustelo**

WHITE ENSIGN MODELS DKM ANTI-AIRCRAFT WEAPONS
1:400 Scale

If you don't get the WEM 1:400 scale ship set, WEM PE 4006 provides five types of light AA weapons on this fret. There are fourteen single 20mm guns, 2cm Flak C30, each of which consists of three parts – pedestal with handwheel, barrel with folding shoulder rests and ammunition clip. There are ten twin 20mm mounts, each of which has seven parts: barrels with folding shoulder rests, pedestal, two ammunition clips, half shield, full shield and base plate. There are six of the quad 20mm mounts, 2cm Flakvierling, each with seven parts, comprising two twin barrels, mount, two shield halves, base with two folding gunner's seats and a separate site/gunner's seat piece. You have to shape the centre part of the mount from resin scrap or plasticard.

■ Sample: **Courtesy of White Ensign Models**
www.whiteensignmodels.com

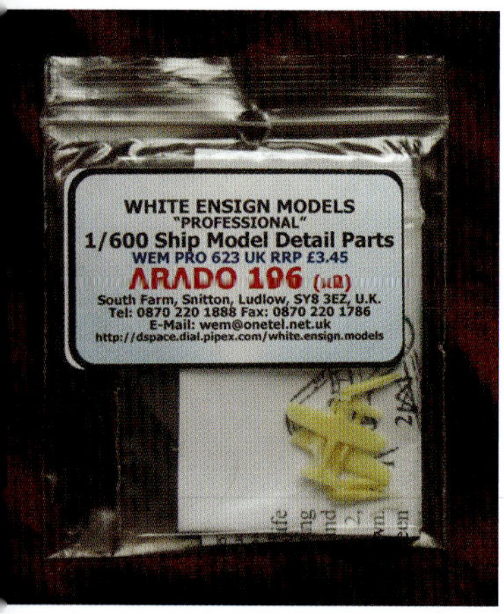

WHITE ENSIGN MODELS ARADO 196 — 1:600 Scale

If you don't like the rather inaccurate Heinkel floatplanes included in the Airfix model of *Scharnhorst*, White Ensign Models provides two beautiful miniature Arados in their WEM PRO 623 set, with resin aircraft and floats and a brass photo-etch fret with propeller, float struts and aft machine gun.

Sample: **Courtesy of Felix Bustelo**

WHITE ENSIGN MODELS DKM DOORS AND WINDOWS — 1:400 & 1:350 Scales

The 1:400 scale photo-etch brass fret of Kriegsmarine portholes and doors provides a tremendous variety of these overlooked fittings. Each and every one is magnificently relief-etched. The fret contains: 28 doors with open portholes; 2 small doors/hatches; 76 rectangular windows in three different styles; and 127 round portholes in three different styles with rigols. All of the windows and portholes have the solid cover open from the frame so the modeller can feature the porthole open or closed. It will take time to attach these, one at a time, over the featureless holes on the Heller *Gneisenau* or *Scharnhorst* kits but they will pump up the detail of both kits dramatically. PE 3579 gives you the same treatment but in 1:350 scale. Both the Dragon *Scharnhorst* and much rarer Iron Shipwright *Scharnhorst* kits will benefit from this detail.

Sample: **Courtesy of White Ensign Models**
www.whiteensignmodels.com

WHITE ENSIGN MODELS DKM DOORS AND WINDOWS — 1:700 Scale

This set is not a smaller version of the White Ensign Models DKM Doors and Windows set in 1:400 scale. All doors and windows are in closed position with WEM PE 764 and concentrates on doors. There are 140 doors in five different patterns and 34 square windows. All of these small brass parts are beautifully executed with fine relief-etching.

Sample: **Courtesy of White Ensign Models**
www.whiteensignmodels.com

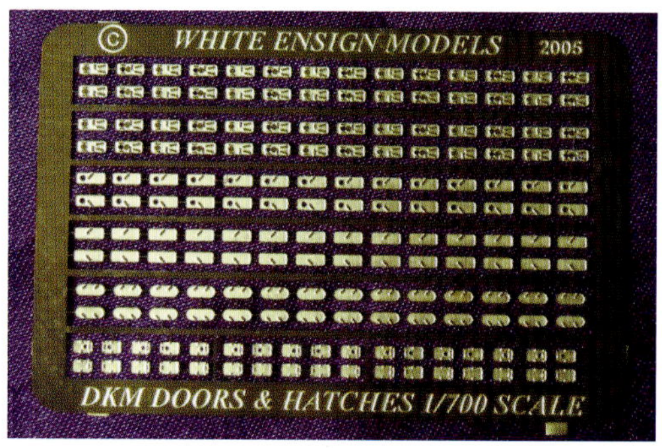

WHITE ENSIGN MODELS CABLE REELS — 1:700, 1/400 & 1/350 Scales

Every navy had its own designs for cable reels and these fittings were very common on any warship. White Ensign Models provides individual cable reel sets in 1:700 PE 786, 1:400 PE 4013 and 1:350 PE 35107. Each set provides nine different Kriegsmarine reel designs. Quantities vary by scale with the most parts provided in the 1:700 scale

Sample: **Courtesy of White Ensign Models**
www.whiteensignmodels.com

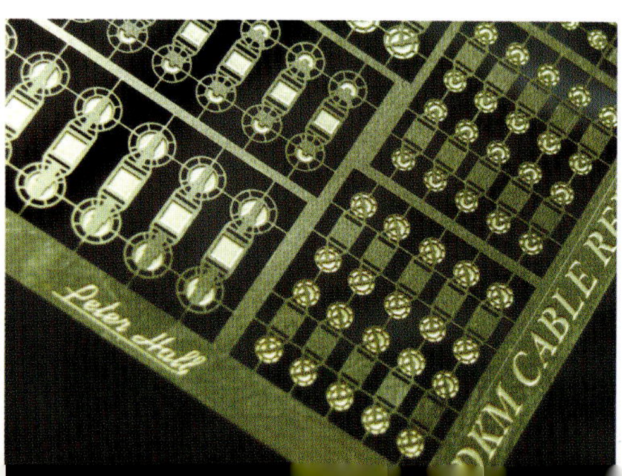

Modelmakers' Showcase

TAMIYA ***GNEISENAU*** 1:700 scale

By CHRISTOPH MENTZEL

Christoph writes: Here are some photographs of my completed model of the 'old' Tamiya 1:700 scale plastic kit. To give the kit more detail and more accurate appearance I used the Flyhawk super-detail set, which also included brass barrels for the main and smaller guns. My kit depicts the battleship *Gneisenau* in the summer of 1940 in Trondheimfjord, before the ship returned to Kiel on 25 July 1940 for repairs which lasted until November 1940. During the time in Trondheim, *Gneisenau*'s original 50/51 camouflage scheme was modified with some darker patches on her hull and superstructure. These darker patches were then over-painted at the time of her return to Kiel. The upperworks returned to her original lighter grey and only the hull was painted with some irregular patches of different shades. My diorama also includes another unit that took part in the earlier Operation Juno, the German supply ship *Dithmarschen*, from a HP Models resin kit.

HP MODELS *GNEISENAU* 1:700 scale

By CHRISTOPH MENTZEL

Christoph writes: After the Channel break-through in February 1942 *Gneisenau* was seriously damaged during the night of 26/27 February 1942 at Kiel in a night attack by the RAF. The whole forecastle, the bow and the first main gun turret were destroyed after a bomb hit. As a result a decision was made to rebuild and reconstruct the *Gneisenau* at Gotenhafen. The three triple 28cm (11in) turrets were to be replaced with three twin 38cm turrets (15in), which now became available because of the cancellation of the 'O' class battlecruisers they were intended for. At the same time the stem and the forecastle would be replaced by a new extended bow

which would give the *Gneisenau* a greater length overall. *Gneisenau* decommissioned on 1 July 1942 at Gotenhafen and the reconstruction began immediately with the removal of the main guns and their turrets. After the disappointing Operation Regenbogen in December 1942 Adolf Hitler ordered the decommissioning and scrap-

ping of all major German surface ships, and this decision halted all work on the reconstruction of the *Gneisenau*.

HP Models has released a new kit of the reconstructed *Gneisenau* as she would have appeared in 1943/44. Before this the only available kits of this class in 1:700 scale were the *Gneisenau* and *Scharnhorst* from Tamiya, which were generally good kits with the potential to build a better model with a bit of additional work. Overall the HP resin kit makes a nice impression from the beginning. It has fine casting with a lot of details on the surface. Many of its parts, especially of the superstructure, are revised components of the Tamiya kit, so the quality is as good and sometimes better than the equivalent Tamiya pieces (the new 38cm guns have metal barrels, for example). The dimensions are correct for the official plan for the ship reconstruction.

I have built my kit as a part of a diorama, showing a typical mooring in a Norwegian fjord at the end of 1943 or early 1944. *Gneisenau* wears a typical German

splinter camouflage as the *Tirpitz* and other German ships had worn it at that time in Norway. Several typical small combatants of the Kriegsmarine are also shown in my diorama: an M-boat, R-boats, *Kriegsfischkutter* and *Vorpostenboot*, all of which are from HP Models, a small *Kriegstransporter* from Delphis, and several scratch items.

SCRATCH-BUILT *SCHARNHORST* 1:300 scale

By MARIUS CHESCU

Marius writes: I just recently completed this model. You won't believe it but it took me only 700 hours! Unfortunately my documentation was not very detailed, so I had to improvise little things. I made the hull plank-on-frame and covered by fibre-glass. I used wood, aluminum sheet and plastic sheet to make the superstructures and deck details. The deck is of pear wood. The rust is real rust and not paint.

DRAGON *SCHARNHORST* 1:350 scale

By KOSTAS KATSEAS

Kostas Katseas is a master at injecting drama into his builds of commercially available injected plastic models. His waterline setting of the Dragon 1:350 scale *Scharnhorst* brings this wonderful kit to life.

TAMIYA *SCHARNHORST* 1:700 scale By KOSTAS KATSEAS

This is *Scharnhorst* in 1939 built by Kostas Katseas in 1:700 scale from the Tamiya kit and with the Flyhawk detail set. She is shown with the crew manning the side to honour the returning *U47*, after the sinking of the *Royal Oak* at Scapa Flow. When you look at the detail packed into this diorama, it is a tribute not only to Kostas's consummate skill, but also to what can still be accomplished with the older Tamiya kits of *Scharnhorst* and *Gneisenau*.

HELLER *SCHARNHORST* 1:400 Scale

By JAN VARGA

Jan Varga writes: This is my attempt to model *Scharnhorst* as she looked in the summer of 1943. I made a lot of scratch-built modifications to the original kit. The main problem of the kit is an incorrect position of the aft turret. But this I left as it is, because correction would require completion of a new deck from scratch. I used White Ensign Models *Scharnhorst* photo-etch set plus 1:350 scale crew figures from Eduard. For a paint job and weathering I tried to follow the method described in the Polish magazine *Modelarstwo Okretowe* using Vallejo Model colour airbrushed on a surface primed by black enamel. The baseboard for the model is a cheap IKEA shelf, covered by Vallejo Atlantic Water Effect gel. Waves are sculpted using a small coffee spoon.

TAMIYA *SCHARNHORST* 1:700 scale

By FRANCISCO ROMERO PEDROSA

This is Fran's build of Tamiya's 1:700 scale *Scharnhorst* and a 1936A type destroyer. He built this diorama in January 2003 using the ABER ship railing and the Tom's Modelworks figures in 1:700 scale.

SCRATCH-BUILT ***SCHARNHORST*** 1:200 SCALE

By JOHN R HAYNES

A fine example of what can be achieved at a relatively large scale by a first class professional modelmaker. It was an earlier model of this ship which launched John's career in 1977: at 1:150 scale, it took four years to build and is now displayed aboard the preserved cruiser *Belfast*, the last survivor of the squadron that sank the German battleship.

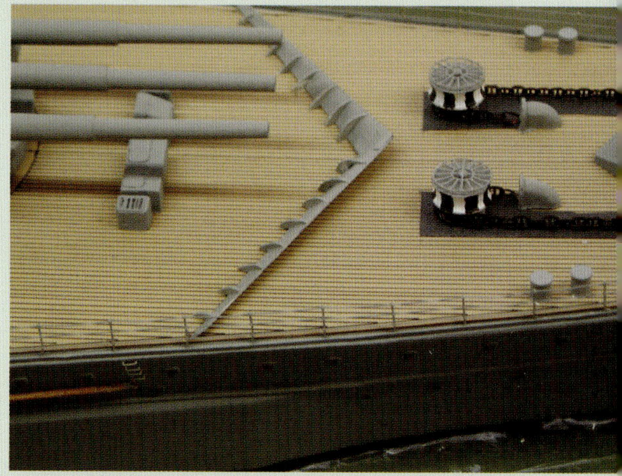

DRAGON *SCHARNHORST* 1:350 scale

By PHIL REEDER

Dragon's 1:350 scale *Scharnhorst* kit is the newest and the best commercially available model of the ship. This is Phil Reeder's build of the exceptional kit.

HELLER/AIRFIX *SCHARNHORST* 1:400 Scale

By PHIL REEDER

Heller is no longer in operation, but Airfix acquired the right to produce the Heller kits. There is no difference between the original Heller 1:400 scale release and that packaged by Airfix. Phil Reeder built the Airfix release of the 1:400 scale *Scharnhorst*. He further detailed it with White Ensign Models photo-etch and some scratch building.

TAMIYA *SCHARNHORST* 1:700 scale

By TED BUNN

Eduard and Tom's Modelworks brass photo-etch was added to during the build. Rigging is from stretched sprue and nylon paint brush bristles. Bow waves were made from Milliput clay and the sea from Liquidex acrylic medium. The Tamiya *Scharnhorst* was the first model Ted built back in 2002 after a 23-year hiatus from the hobby. Ted has a Kriegsmarine cap band, Iron Cross and uniform eagle with swastika that belonged to his wife's uncle. His name was Hans Klaussner and he was an electrician on the *Z29*, a destroyer with a strong connection to the *Scharnhorst*. During the Channel Dash, after *Scharnhorst* had gone dead in the water from striking a mine, it was *Z29* that was ordered to close *Scharnhorst* and embark the fleet commander, Vice Admiral Ciliax. *Z29* was also with *Scharnhorst* in Norway and was one of the escorting destroyers for *Scharnhorst* and *Tirpitz* in Operation Eisberg resulting in the shelling of Spitsbergen's installations. Finally, *Z29* was one of five destroyers escorting *Scharnhorst* on her last sortie, which were detached to return to Norway before *Scharnhorst*'s final encounter with HMS *Duke of York*.

AIRFIX *SCHARNHORST* 1:600 Scale

By BERND VILLHAUER

With this build of the Airfix 1:600 scale *Scharnhorst*, one can see the greatly different appearance of the ship as built, with an almost straight stem, flat funnel top and Heinkel biplanes, compared to the more familiar image with Atlantik bow, raked funnel cap and Arado monoplanes. Only this kit portrays the as-built fit. Bernd Villhaur built it and further detailed it with White Ensign brass photo-etched parts and Colourcoat Paints. He also scratch-built additional detail.

HELLER *SCHARNHORST* 1:400 scale

By FRANCISCO ROMERO PEDROSA

Francisco writes: Ten years ago, when I began in this hobby, I built a small diorama in 1:700 scale of the German battleship *Scharnhorst* at Gotenhafen. The *Scharnhorst* was a ship of an interesting design that had a tragic but impressive history; and it was my favourite warship. After some years I decided to build it again in a larger scale. I do not have experience with models of resin so I obtained a 1:400 scale Heller kit. Although it is an old and very basic kit, it was the best thing in plastic. To transform the ship to a 1943 version, I needed a lot of a lot of additional information. I used White Ensign Models PE4002 photo-etch. The funnel also needed a lot of work because there is practically no detail.

With patience and with sufficient information, these parts can be detailed to the limit of one's ability. New boats, exterior fans and the platforms for the anti-aircraft guns were added. The artillery, just like the remainder of the components of the kit, comes with very few details and needs some extra work. The gun barrels are of such poor quality and unrealistic, that I replaced them with barrels from RBM. The improvement is evident. The platforms for the quadruple Flak guns were made from scratch. These were installed during the ship's time in Brest. The admiral's bridge and the radar platform are improved with photo-etch parts and wire of different sizes. New platforms need to be constructed with inclined ladder added since they do not exist in the kit. A lot of work had to be done to the armoured bridge with new windows added as well as the lookout positions. The bow needed some attention too. To show the open hangar some cutting was required on the high part and a new roof fold was manufactured; the catapult is placed on top of the hangar. The photo-etch version was substituted for unrealistic plastic version. Extra details were added, such as the bottles of compressed air.

The aft mast needed to be reconstructed, with new platforms added to match the fit in 1943. For the painting I used the plans of Antonio Bononi. At the beginning of 1943 the ship carried a camouflage applied for Operation Paderborn, which I personally

believe to be the most characteristic one for this ship. In the diorama there are three aircraft, although the Heller kit only comes with one. The other two Ar 196s are from L'Arsenal and are a lot better detailed so I modified them to be able to show their crew. I cut the canopy off completely and drilled out the cockpit, cut down some figures and glued them in. New canopies were fabricated with a sheet of thin transparent plastic. In this diorama I tried to represent a realistic scene. Many modelmakers place their crew figures randomly, apparently with the only intention to fill up spaces. In this diorama there are almost 200 figures and I studied scores of photos of many real ships to give the impression of specific activities such as provisioning, maintenance, cleaning etc. The figures are L´Arsenal 1:350 scale figures cut down somewhat so they appear more in scale. I took extra care to make sure the figures were not randomly placed but all appear to be doing a specific task. The figures help focus your attention on the details of the diorama.

In port there were three kinds of trains to facilitate the transportation of equipment and cargo, therefore I felt obliged to build a train. This would also help me to give a little life to the diorama. Among the models of surviving steam engines of that era, the one that gets the most attention from model-makers is known as Kriegslokomotive BR 52. It was painted in interesting camouflage patterns, so I chose it to accompany my *Scharnhorst*. I found plans on the internet that I reduced to 1:400 scale and began construction. With a little bit of imagination an acceptable engine can be built (the wheels, for example, are from some photo-etch cable reels). The four freight cars that accompany the locomotive carry boxes of munitions and torpedoes. In the first freight car I placed a Flak 38 AA mounting.

Gneisenau as commissioned 1938

The two tones of grey are the standard peacetime Kriegesmarine colour scheme.

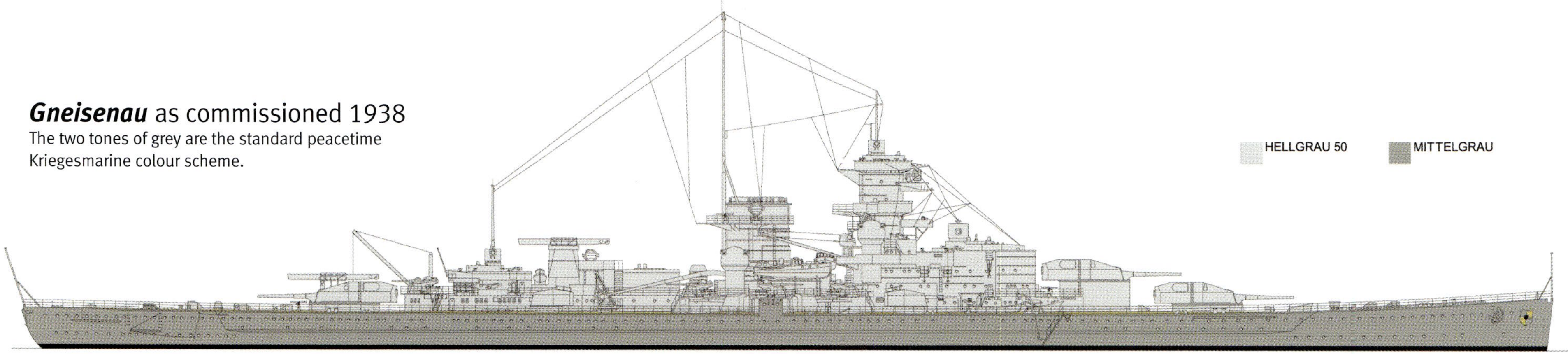

Gneisenau in Baltic camouflage early 1941

Early in the war red turret tops were used as an aerial recognition system when in range of Luftwaffe aircraft, often combined with prominent swastikas painted fore and aft. Yellow and dark grey were also used in lieu of red in some instances. The black and white zig-zag markings were only used in the Baltic.

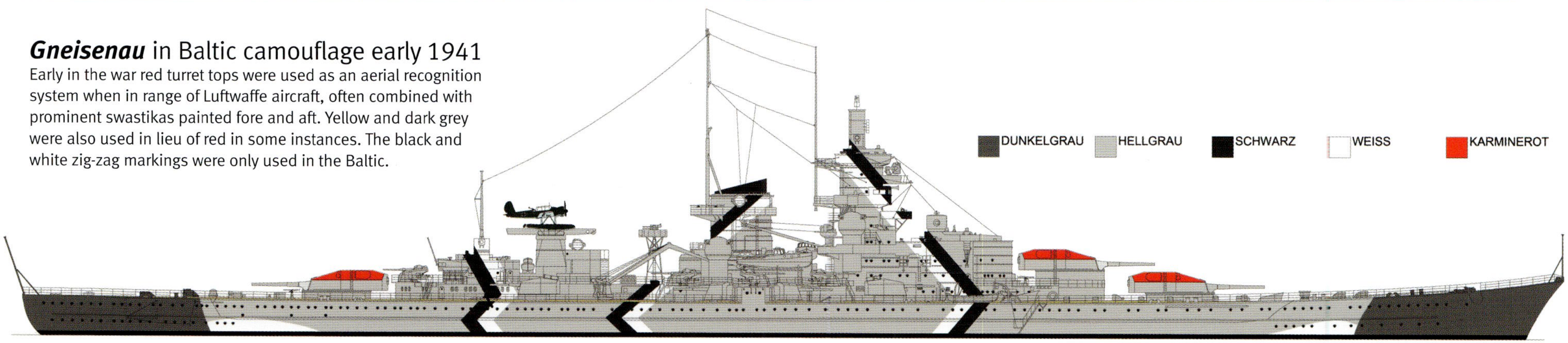

Scharnhorst **in April 1940 for Operation Weserübung**

The ship probably carried yellow turret tops during the first phase of the invasion of Norway; they were repainted red later in the operation.

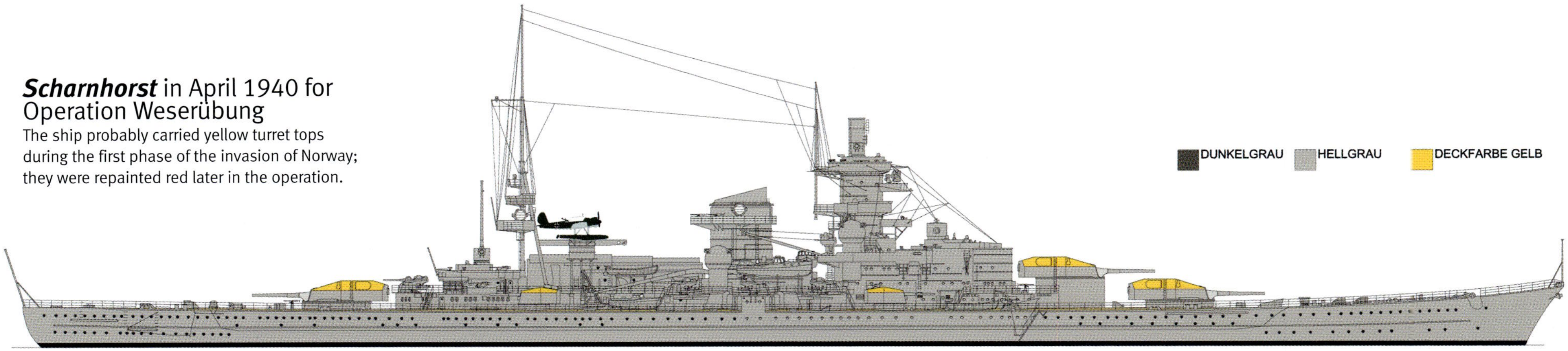

Scharnhorst **in late 1940**

The Baltic stripes amidships are the remnants of a more elaborate pattern of black and white broken stripes carried earlier in the year.

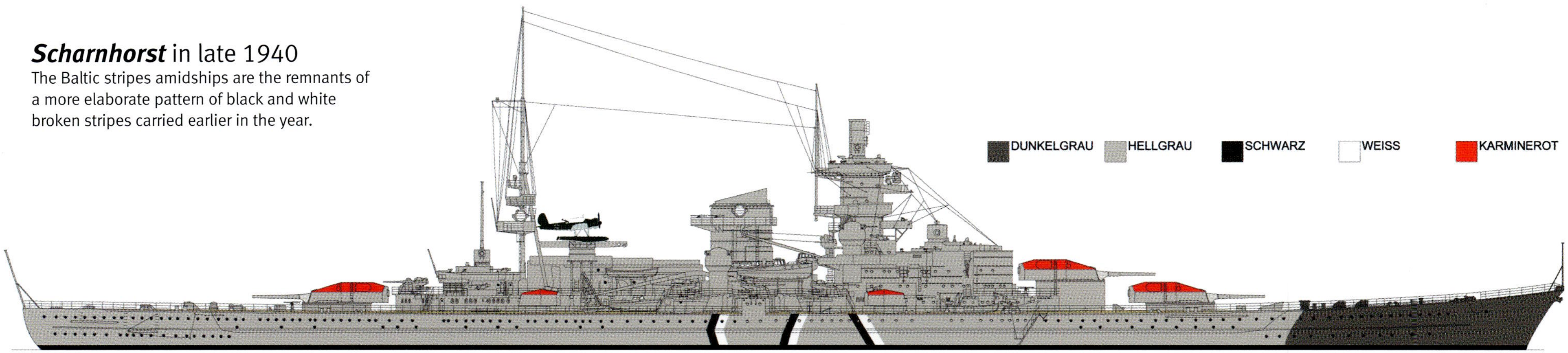

Scharnhorst **in September 1943**

This is the scheme carried by the ship during the Spitsbergen raid.

Appearance

STRUCTURAL AND EQUIPMENT CHANGES

The *Scharnhorst* class was designed with a straight stem and a flat-topped funnel, the same design features found on the *Deutschland* class. This created a family appearance, which was thought to confuse an enemy over the type of warship that was sighted. During the trials of both *Scharnhorst* and *Gneisenau* it was discovered that the low sheer, straight bow and high speed of the design made for an extremely wet forecastle and the flat stack allowed exhaust fumes to be drafted into the superstructure. As a result the ships were given a clipper or Atlantik bow and a funnel cap, and this became the new family appearance, as the same changes were added to the *Bismarck* class, *Hipper* class before the war, and the surviving *Deutschland* class during the war. *Gneisenau* was taken in hand for these modifications in January 1939 and *Scharnhorst* in June 1939, the latter's refit being completed just after the outbreak of the Second World War.

As built, the pair had a traditional anchor arrangement with anchor chain running across the forecastle from the cable holders to deck hawse and out through the hull in hawse pipes. The anchors were housed in their hawse about three-quarters up the hull with two anchors to port and one to starboard. When *Gneisenau* was first fitted with the Atlantik bow in May 1939 the original anchor arrangement was retained, but it was soon discovered that the anchors mounted at that location still created water resistance and excessive spray in heavy seas. As a consequence, the anchors were moved higher on the new clipper bow. The three anchors were reduced to two, with one on each side, mounted at the deck-edge in new channels or chutes, inclined from the forecastle out over the upper hull. The old hull hawse fittings were removed, and the pipes sealed. The lesson was learnt before the *Scharnhorst*'s refit was complete so that ship went straight to the new arrangement, but added a third anchor at the stemhead.

Initially *Scharnhorst* and *Gneisenau* were very close in appearance and extremely difficult to distinguish one from the other, although *Gneisensau* carried a short foremast on the maintop director until the second stage of her bow reconstruction. When *Scharnhorst* went in for refit for the Atlantik bow in June 1939 she also received another modification, not received by *Gneisenau*, which would change her profile dramatically and thereafter make it easy to identify each ship. As built, the pair had a mainmast stepped aft of the funnel but in front of the hangar. *Gneisenau* retained this arrangement when her bow was rebuilt, but since *Scharnhorst* received her refit five months later, more thought had been given to further improvements of the class. *Scharnhorst* was given a longer and larger hangar to accommodate two aircraft inside. The original mainmast was replaced by a new light tripod mainmast with platforms placed aft of the hangar, making the silhouette of *Scharnhorst* very different from that of *Gneisenau*. On both ships the admiral's bridge on the superstructure tower was originally open to the elements. In trials the open bridge proved to be less than conducive for efficiency. *Gneisenau* had her admiral's bridge enclosed in January 1940.

Below: *Gneisenau* after the initial refit with her new funnel cap, probably about August 1938, showing the first stage of the modified Atlantik bow. Note that the anchor is still in its original position, as is the foremast. This funnel cap fitted flush with the outer casing of the stack, incorporating all the steampipes originally visible around the rim; it was quickly replaced by a narrower version when the anchor hawse was modified.

Above, right: A close-up of *Scharnhorst*'s funnel as completed. Note the small steampipes around the funnel rim.

Above, left: A view aft from the bridge of *Gneisenau*'s modified funnel cap in April 1940. The first cap was the full width of the funnel, but the steampipes around the rim re-emerged when this narrower cap was fitted.

Right: Contrasting views of *Gneisenau*'s bridge structure: as completed and with the admiral's bridge enclosed – the iced-up forecastle seen in the second picture, dating from the harsh winter of 1940/41, demonstrates why it was a good idea.

The fall of 1939 also saw the introduction of radar arrays atop the forward superstructure director. The FuMO 22 'mattress' aerial was installed on *Gneisenau* in October 1939 and on *Scharnhorst* in December 1939. Also, at some point late in 1939, the bow crests were removed and stored ashore. Both *Scharnhorst* and *Gneisenau* had the crests after they were completed with their Atlantik bows but photographs of both ships taken in January 1940 show the crests were no longer present at that point in time.

There were comparatively few changes in 1940. Both ships received a degaussing cable, which ran along the top of the armour belt and was designed to reduce the magnetic signature of the ship as a defence against magnetic mines. Another change was more obvious. Both ships originally had a catapult on the crown of 'C' turret. The aircraft was vulnerable to blast damage, but it was anticipated that this position posed a danger to the gun turret itself. Designers correctly foresaw that the aviation fuel and lubricants represented a very high risk of fire that might conceivably threaten the operation of the turret. During the Battle of Savo Island in August 1942 the US Navy lost three heavy cruisers in large measure because of uncontrollable fires started in their amidship hangars and fed by the aviation fuel and lubricants. The *Scharnhorst* and *Gneisenau* landed their turret catapults and associated aircraft derrick early in 1940.

Even fewer changes were made in 1941, involving only anti-aircraft guns and radar. The single 20mm C30 light AA guns were replaced by 20mm C38 models. The FuMO 22 radar on the forward superstructure was replaced with a newer array, the FuMO 27, as well as adding a second FuMO 22 array on top of the aft fire control director. *Scharnhorst* received two triple 533mm (21in) torpedo mounts, which had been landed from the light cruiser *Nürnberg* and were placed amidships with one mount on each side. [During early operations against convoys, the length of time it took to sink AMCs like the *Jervis Bay* allowed the bulk of the merchant ships to escape; the torpedoes were designed to finish off such resistance quickly.] The navigation bridge wings were also shortened on *Scharnhorst*. *Gneisenau* received a single quadruple 20mm AA gun mount on a light lattice tower amidships. Most of the year was spent at Brest under camouflage nets and under constant attack by the RAF. It was clear that their anti-aircraft armament needed strengthening beyond just replacing single 20mm guns with a newer pattern.

Above: The back of the bridge of the *Scharnhorst* probably taken at the turn of 1939/40. Although the director is turned away from the camera, careful inspection reveals the FuMO 22 radar aerial on its front face. Note how the foremast was stepped on a small bracket abaft of the admiral's bridge platform.

SUMMARY OF *SCHARNHORST* MODIFICATIONS

Date	AA Modification	Other modification
June 1939		Hull lengthened, clipper Atlantik cutwater added, raked funnel cap added, hangar enlarged, pole mainmast landed and new light tripod mainmast added aft of hangar, modified anchor arrangement with bower anchors reduced from three to two, but third at stemhead
December 1939		FuMo 22 radar fitted to foretop director, bow crests and stern eagle landed for storage ashore
March 1940		Degaussing cable added and 'C' turret catapult and derrick landed
1941	Single 20mm C30 light AA guns were replaced by 20mm C38 models	FuMo 22 radar replaced by FuMO 27, FuMO 22 radar added to the top of aft director, two triple 533mm torpedo mounts fitted, navigation wings shortened
January 1942	Five 20mm Vierling quadruple gun mounts added	Admiral's bridge enclosed
March 1942	Two 20mm Vierling quadruple mounts landed	
December 1942	Three 20mm Vierling quadruple mounts added	Replaced FuMO 27 radar with FuMO 26 radar
March – September 1943		Various radar detection equipment fitted
November 1943		New FuMO 26 radar forward and FuMO 61 to foremast

Above: An Ar 196 floatplane being hoisted aboard *Gneisenau*. The design of the cranes in the two ships was distinctly different, not just in the shape of the jibs but also in the way they were constructed. Note the 'duckboard' with its prominent treads running up the outside of the jib.

Opposite, middle: The aircraft arrangements of the two ships were very different from early on in their careers. *Gneisenau* never had the larger hangar fitted to her sister and in 1940 the smaller one was also removed, leaving the catapult perched on a prominent pedestal. Note the position of the searchlights, which were mounted on towers that could be raised when required, although the catapult could not be used when they were at full extension.

Above: Details of the funnel platform and mainmast step of *Gneisenau*. The photo was probably taken around November 1939 when working up after repairs to the damage received off Norway. The new FuMO 22 aerial can just be seen above the plume of funnel smoke. The mast was usually painted black, presumably because it would be soot-blackened in any case.

Below: *Scharnhorst* alongside at Wilhelshaven in March 1940. In the background the catapult is being removed from 'C' turret, with the help of a floating crane outboard of the ship. In the foreground what appear to be engineering ratings are being instructed in aspects of the ship's main machinery.

Above, left: The quarterdeck of *Gneisenau* towards the end of 1939. The catapult, offset to starboard, is still fitted atop 'C' turret, but it would be removed at Deutsche Werke in February 1940.

Above, right: One minor difference between the two sisters was the shape of the jib of the aircraft-handling crane. *Scharnhorst*'s, seen here, was curved at the end, whereas the *Gneisenau*'s crane had a distinct angle.

Left: The after superstructure of *Gneisenau* at Brest in March 1941 after her long Atlantic raid (Operation Berlin). Note that the turret tops have been painted, probably red, for the approach to Luftwaffe-controlled airspace. The platform on the light latticework pylon abaft the funnel carried a Vierling 20mm mount. The tall searchlight towers either side of the catapult are at their full height; they were fitted to retract when the catapult was in use.

SUMMARY OF *GNEISENAU* MODIFICATIONS

Date	AA Modification	Other modification
January 1939		Hull lengthened, clipper Atlantik cutwater added, raked funnel cap added
May 1939		Modified anchor arrangement – reduction from three to two bow anchors
October 1939		Fitted FuMO 22 radar on top of forward director
January 1940		Admiral's bridge enclosed, bow crests and stern eagle landed for storage at some time before January 1940
Early 1940		Degaussing cable added and 'C' turret catapult and derrick landed
1941	One 20mm Vierling quadruple mount added; single 20mm C30 light AA guns were replaced by 20mm C38 models	FuMo 22 radar replaced by FuMO 27, FuMO 22 radar added to the top of aft director
1941-January 1942 at Brest	Three 20mm Vierling quadruple gun mounts and additional two single 20mm guns fitted	Hangar lengthened and catapult removed from pedestal and added to interior of hangar, two triple 533mm torpedo mounts added
July 1942		Main gun turrets removed for planned replacement with twin 15in gun turrets; project cancelled and ship lacked main gun turrets for rest of war

Right and below: Although *Scharnhorst* had a larger hangar, stowing the floatplanes was still a complicated and time-consuming manoeuvre, involving the manual folding of the wings and gently lowering the aircraft into the hangar.

In January 1942 this took the form of multiple quadruple 20mm gun mounts. Known as the Vierling, these four-barrelled mounts greatly increased the short-range anti-aircraft defence.

In January 1942 *Gneisenau* received three Vierling and an additional two single 20mm guns. The quadruple mounts were placed on the crown of 'B' turret, the funnel platform and on the after superstructure. For *Gneisenau* the catapult had been removed when the hangar was enlarged with a new catapult installed inside the hangar. A new mainmast along the lines of that carried by *Scharnhorst* was due to be installed aft of the hangar; it was actually at Brest awaiting fitting, but the ship left for the Channel Dash before it could be installed. During this period *Gneisenau* was fitted with triple torpedo mounts, like those carried by *Scharnhorst* since 1941. For *Scharnhorst* Vierling were fitted on the funnel platform, on the crowns of the forward 5.9in gun turrets on each side, on both sides of the hangar and on the crown of 'B' turret. *Scharnhorst* also had her admiral's bridge finally enclosed.

After Operation Cerberus, the Channel Dash, the pair finally separated. When the forward section of *Gneisenau* was badly damaged in an air attack on 26/27 February, it was decided to rebuild the ship. The class had been designed to allow the triple 11in gun turrets to be replaced with twin 15in gun turrets and *Gneisenau* was to receive this treatment. The work was to be done at Gotenhafen in the Baltic, beyond the easy reach of the RAF. The shattered forward section was removed with plans to reconstruct it with an even longer bow. The ship would also have received a modified hangar and catapult, and a tripod mast aft would have finally replaced the mainmast on the funnel platform. However, the work never progressed very far, and for reasons pointed out in the Careers section, was abandoned at the end of 1942.

Scharnhorst received minor changes after her return from France. In March 1942 two of her Vierling were landed but she received three Vierling in December 1942 while still in Germany. A FuMO 26 array replaced the FuMO 27 radar forward. After she arrived in Norway in March 1943, major alterations ceased, with the later changes confined to additional radar detector equipment in March and September. Her last modifications came in November 1943 with a new FuMO 26 radar array on the forward director and a FuMO 61 array added to the foremast.

CAMOUFLAGE AND PAINT SCHEMES

For the most active portion of their combat careers *Scharnhorst* and *Gneisenau* wore

Above: *Scharnhorst* in the autumn of 1942 at Gotenhafen where repairs were carried out following the damage sustained during the Channel Dash. By this stage of the war, the bold black and white zig-zag stripes usual in the Baltic had been discontinued.

Below: The mainmast of *Scharnhorst* while the ship was in dock at Brest. Note the makeshift camouflage on the mast platforms and the top of the 4.1in gun mounting.

Right: *Gneisenau* during Operation Juno in June 1940. Following usual Kriegsmarine procedure, the turret top is painted, part of the measures taken to make German ships instantly identifiable to the Luftwaffe. These also included large swastikas painted on the weather deck fore and aft. During the invasion of Norway, the turret roof colour seems to have been yellow, but then was replaced by red; dark grey is also known in some cases, and for the Channel Dash the major ships seem to have painted their turret tops blue.

the same pre-war paint scheme. Sometimes confused as an overall single-tone grey scheme, in reality the pre-war paint scheme was two tones of grey. The superstructure was a pale grey (*Hellgrau*) and a darker grey (*Mittelgrau*) for the hull. The pair wore this scheme in their 1940 and early 1941 Atlantic raids. One exception to this constant scheme was the addition of black and white diagonal stripes and a dark bow and stern while working up in the Baltic in 1940 and 1941. In common with all heavy ships training in the Baltic, the black and white stripes were painted over prior to any venture into the Atlantic. For Operation Berlin from January to March 1941, the ships had the crowns of the 11in and 5.9in turrets painted yellow. This was for aerial recognition by Luftwaffe pilots, who in common with other ground-based aviators had difficulty distinguishing one ship from another. It was common for ships in the Baltic to have their turret crowns painted wine red (*Karminerot*) for the same purpose. It is unlikely that the battleships had their turret crowns painted yellow at the start of the sortie, as the only aircraft that they would encounter in the mid-Atlantic would be hostile. Far more logical is that they were painted yellow near the end of the sortie

SUMMARY OF *SCHARNHORST* CAMOUFLAGE PATTERNS

Date	Location	Pattern
January 1939–early 1942		Standard, but modified as following: Light grey (*Hellgrau*) superstructure, Medium grey (*Mittelgrau*) hull
April 1940	Operation Weserübung, Norway	As above but possibly yellow turret crowns since February 1940
1940–1941	Baltic work-up periods	Standard, with diagonal black and white stripes added
March 1941	Operation Berlin, Atlantic raid	Standard, with yellow turret crowns
February 1942–?	Operation Cerberus, the Channel Dash	Overall light grey but with two-level dark grey panels most of the length of the hull, light grey superstructure with dark grey mottled pattern, blue turret crowns
January 1943–summer 1943	Baltic and Norway	Dark grey (*Dunkelgrau*) superstructure and bulk of hull, pale grey bow and stern, false profile pattern
Summer 1943–December 1943	Norwegian Fjord	Disruptive Pattern Multiple greys in combination of curves and angles for superstructure and bulk of hull, light grey bow and stern

SUMMARY OF *GNEISENAU* CAMOUFLAGE PATTERNS

Date	Location	Pattern
May 1938–early 1942		Standard, but modified as following: Light grey (*Hellgrau*) superstructure, Medium grey (*Mittelgrau*) hull
April 1940	Operation Weserübung, Norway	As above but possibly yellow turret crowns since February 1940
1940–1941	Baltic work-up periods	Standard, with diagonal black and white stripes added
March 1941	Operation Berlin, Atlantic raid	Standard, with yellow turret crowns
February 1942–?	Operation Cerberus, the Channel Dash	Overall light grey but with two-level dark grey panels most of the length of the hull, light grey superstructure with dark grey mottled pattern, blue turret crowns

after the ships had turned towards Brest in order to prevent fratricide from their brethren of the Luftwaffe. *Bismarck* did the same thing in her May 1941 sortie and over-painted the turret crowns yellow only after she changed course for Brest. A photograph of *Scharnhorst* taken at the end of March 1941 at the end of Operation Berlin shows the over-painted Baltic stripes starting to appear as the over paint had worn off in the heavy seas of the Atlantic and it does appear that the turret crowns were of a darker shade than the sides.

In early 1942 in preparation for Operation Cerberus, the Channel Dash, both ships, as well as the heavy cruiser *Prinz Eugen* received a one-off scheme used just for this operation. A dark grey to black two-level panel ran almost the length of the hull and dark grey mottling was painted over a light grey ground on the superstructure. Additionally, the turret crowns for the 11in and 5.9in turrets were painted a light to middle blue. (*Warship Pictorial 36: Kriegsmarine Scharnhorst* contains a colour photograph taken aboard *Scharnhorst* during the Dash, showing a close-up of one of her 5.9in turrets with the blue crown.) *Gneisenau* was probably still wearing this scheme on 26 February 1942 when she received the bomb hit which wrecked her bow and 'A' turret and may have still worn it on her voyage to Gotenhafen. Photographs of the wreck of the *Gneisenau* in 1945 seem to show mottling on the superstructure but this may have been due to rust and weathering instead of the retention of the Channel Dash camouflage scheme.

Scharnhorst also went to Gotenhafen for repairs, which were finished in January 1943. She received a new paint scheme in the Baltic prior to steaming west and starting her attempts to get to Norway. This scheme used two highly contrasting colours. The superstructure and most of the hull was painted a very dark grey; the bow and stern were painted a very pale grey (called *Schnellbootweiss*, or S-boat white) presenting a false silhouette designed to fool the viewer that the ship was farther away in distance than it really was. She arrived in Norway in March 1943 with this scheme and still had it at least into June 1943 and probably carried it during the Spitsbergen raid. Later in September 1943 *Scharnhorst* was given a new 'Norwegian' camouflage pattern designed to allow the ship to blend into the rocky fjord environment. Unlike the *Tirpitz*, which carried an angular ray pattern, *Scharnhorst* received a multi-colour pattern of curves. The ship still had a light grey bow and stern but the majority of the hull and the superstructure had a least three colours used in curves and angles designed to break up the ship's silhouette in a manner similar to the Royal Navy Admiralty Disruptive Pattern. The colours were most likely shades of grey but there is no direct evidence as to the actual colours. *Scharnhorst* carried this scheme at the time of her loss off the North Cape.

Above: Shortly after *Gneisenau* arrived at Brest in March 1941, the ship was visited by a delegation from the Imperial Japanese Navy – not yet in the war but keen to pick up whatever combat lessons were available.

Left: *Scharnhorst* with *Tirpitz* astern in September 1943. This was the appearance of the ships during their raid on Spitsbergen.

Above: The midships area of *Gneisenau* covered in camouflage netting while in dry dock in Brest. The upperworks and horizontal surfaces were given some mottled camouflage that looked very ad hoc. The ships carried something similar on their superstructures during the Channel Dash.

Below: A close-up of a twin 3.7cm SK C30 mounting abaft the catapult in *Gneisenau*.

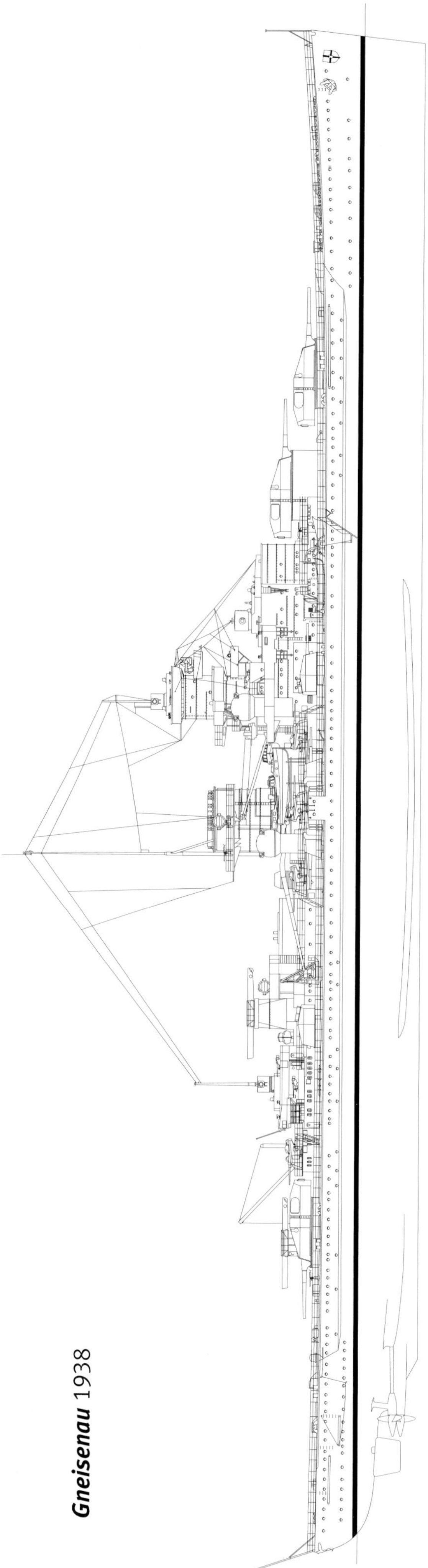

***Gneisenau* 1938**

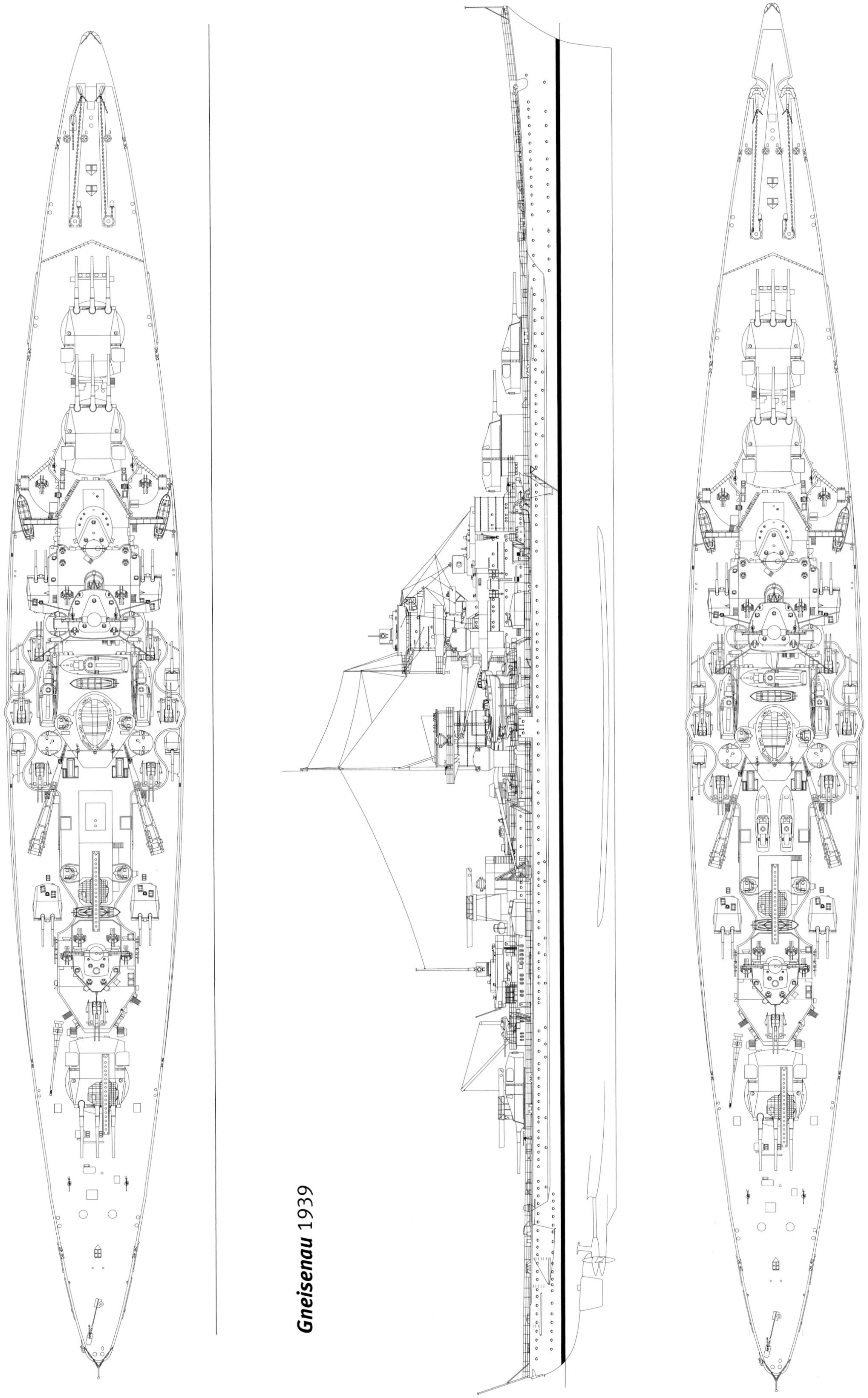

Gneisenau 1939

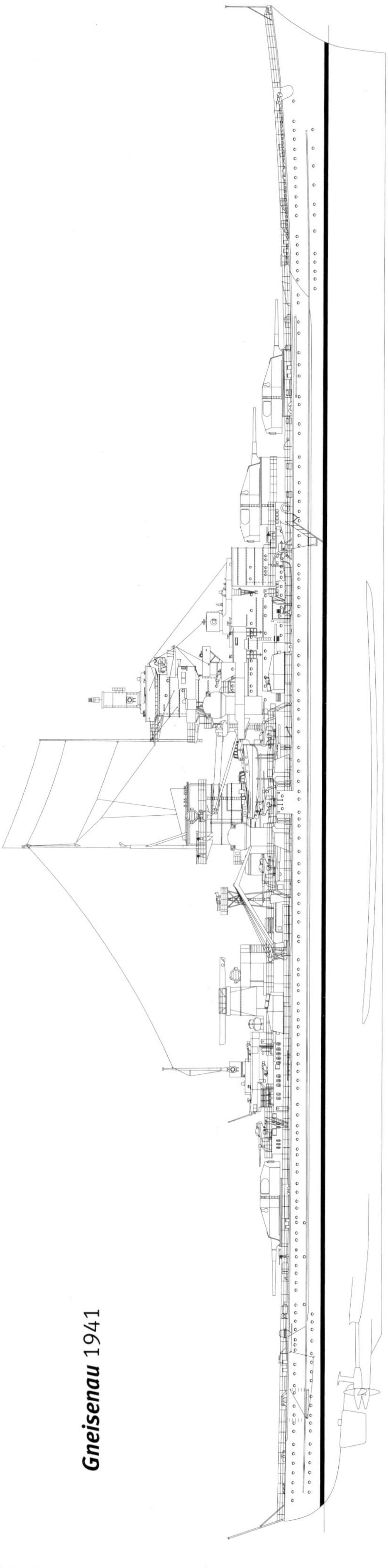

Gneisenau 1941

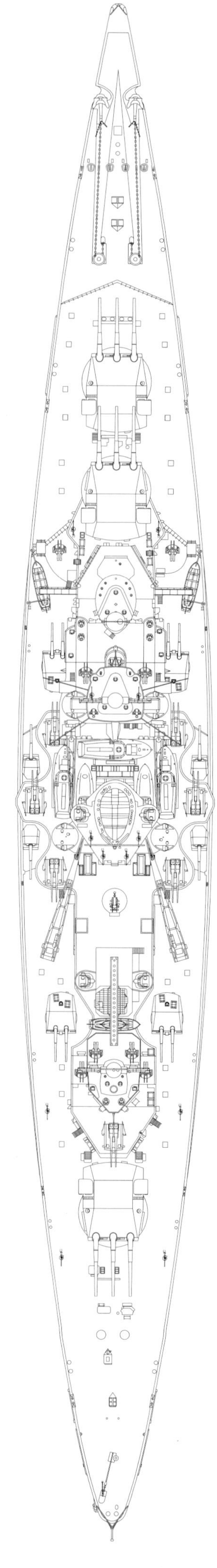

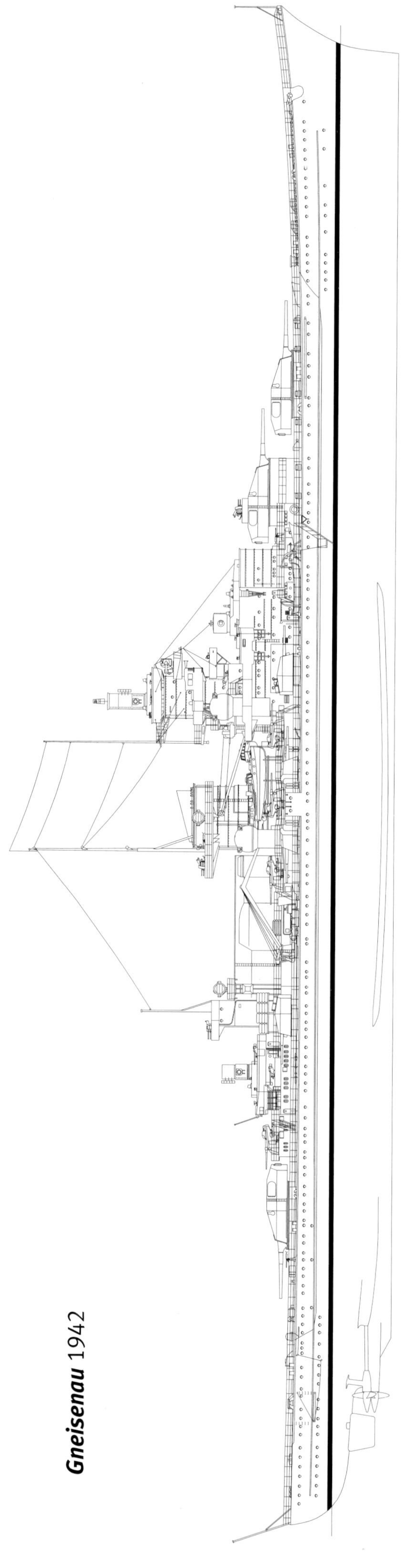

Gneisenau 1942

Gneisenau 1942 – planned conversion

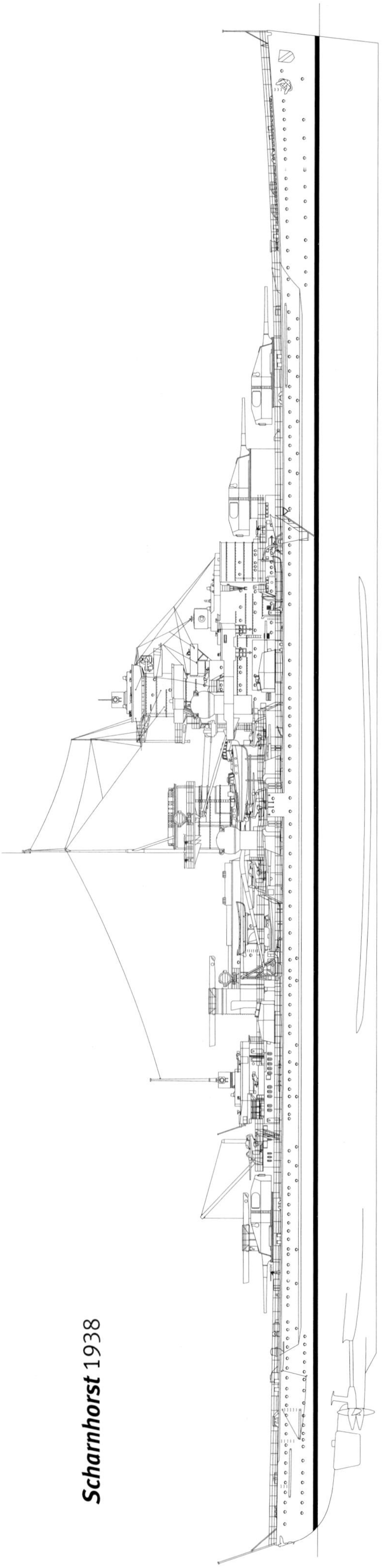

***Scharnhorst* 1938**

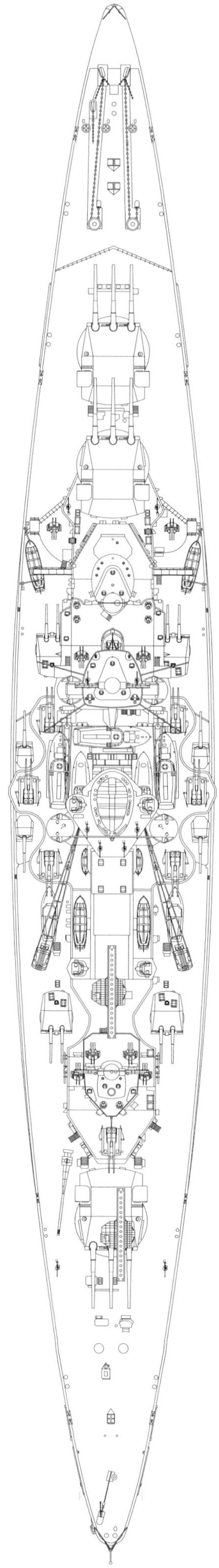

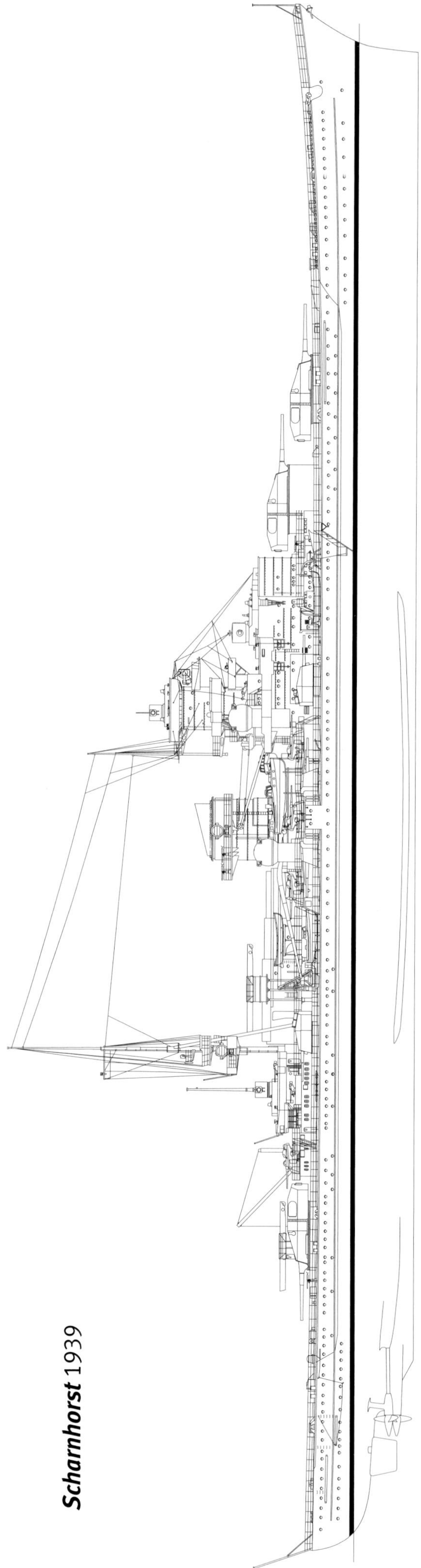

***Scharnhorst* 1939**

Scharnhorst 1943

Selected References

BOOKS

Breyer, Siegfried, *The German Battleship Gneisenau*, Schiffer Publishing Ltd (West Chester, Pennsylvania 1990); originally published as *Schlachtschiff 'Gneisenau'*, Podzun-Pallas-Verlag (Friedberg). Just on *Gneisenau*, this volume is 48 pages, soft bound, largely consisting of photographs of *Gneisenau* with good captions about each. Fairly short text sections on construction history, power plant, armour, armament and planned rebuilding. Also included are profiles for 1938, 1939, 1942 and a plan view and some tables. Only the first 40 pages are on *Gneisenau*, as the last 8 pages cover diverse more modern topics.

Breyer, Siegfried & Skwiot, Mirosław, *German Capital Ships of the Second World War: The Ultimate Photograph Album*, Seaforth Publishing (Barnsley 2012). Large format pictorial book with nearly 100 captioned photos of each ship, plus an introductory chapter on their design evolution.

Federowicz, Przemysław, *Niemieckie Pancerniki typu Scharnhorst, Okrety Swiata/Warships of the World 13*, Okrety Wojenne (Tarnowskie Gory 2001). This hardbound volume of 68 pages is in Polish, but contains numerous photographs, as well as line drawings of armament and aircraft. Both end papers have colour plates with profiles and plans of: *Scharnhorst*, as built, 1938/39; *Gneisenau* 1940; *Scharnhorst* 1941 for Operation Berlin and *Gneisenau* 1941. Also provided are separate 1:400 scale profiles and plans for *Scharnhorst* 1938/39 (as built), *Gneisenau* 1941 and *Scharnhorst* 1943.

Gray, Edwyn, *Hitler's Battleships*, Naval Institute Press (Annapolis, Maryland 1999). This title is an easy-reading, popular history of the operational careers of the battleships down to *Tirpitz*. It contains 195 pages of text and 16 pages of photographs.

Gröner, Erich, *German Warships 1815-1945, Volume One: Major Surface Vessels*, Conway Maritime Press, 1990 for first English edition (London 1990); published in German as *Die deutschen Kriegsshiffe 1815-1945*. Encyclopedic treatment for all German warships for 140 years. Given the scope of the volume, coverage on each class is limited, but includes highlights of construction and careers with tables and profile and plan of *Scharnhorst* 1942 and profile *Gneisenau* 1942 and a small drawing of the *Gneisenau*'s smaller hangar prior to its enlargement at Brest in early 1942.

Koop, Gerhard & Scholke, Klaus-Peter, *Battleships of the Scharnhorst Class*, Greenhill Books (London 1999) & Naval Institute Press (Annapolis, Maryland 1999); originally published in German as *Schlactschiffe der Scharnhorst-Klasse*, Bernard & Graefe Verlag (Bonn, 1991). This volume provides excellent coverage of both ships. Hardbound with 178 pages, the volume presents the best single source for these battleships. Photographs, line drawings, camouflage schemes, everything is included. The text covers the design and operational history, including excerpts from the combat logs of the ships.

Motyka, Mariusz and Nowak, Grzegorz, *The Battleship Scharnhorst*, Kagero Publishing (Lublin, Poland 2010). This volume is very different from all other references in this list. Soft covered and 74 pages in length, the first seven pages contain English text about the design, armament, equipment and history of the ship. What makes this monograph unique is what follows: every aspect of the exterior of the ship is covered in extremely detailed CAD drawings. The last 16 pages of CAD drawings are something of a novelty in that they are rendered in 3D (the volume comes with a set of red/blue 3D glasses). Additionally there are colour plates of *Scharnhorst* in her December 1943 camouflage pattern, as well as one plate on the Arado Ar 196 A3. Two separate large sheets of plans are included, printed both sides in 1:350 scale. In addition to the standard plan and profile, there are sections, isometric views, end-on views and larger size superstructure details.

Schmalenbach, Paul, *German Battlecruisers Scharnhorst and Gneisenau, Warship Profile 33*, Profile Publications Ltd (Windsor UK, 1973). The monograph is 24 pages in length and provides a good mixture of text and photographs. A fold-out four-page colour profile and plan is the centrefold of the volume and presents *Scharnhorst* in 1941 with the false bow and stern markings, although this camouflage scheme was not applied to *Scharnhorst* until January 1943 after repairs from Operation Cerberus.

Trojca, Waldemar, *Pancerniki 'Scharnhorst', 'Gneisenau': Natilus 3*, Wljdawllictwo Militaria (Warsaw 1994). This a soft bound monograph of 40 pages in Polish, providing a chronological history of the ships. It has a good supply of photographs. The centre section provides 1:200 and 1:400 scale drawings of *Scharnhorst*. The back cover has a colour photograph and plates of the ship's crests and aircraft.

Vulliez, Albert & Mordal, Jacques, *Battleship Scharnhorst*, Essential Books (Fair Lawn, New Jersey 1958). Written by two French naval officers, this book is hard bound and of 258 pages. This is a good narrative history of *Scharnhorst*. Of course until after Operation Cerberus, *Scharnhorst* and *Gneisenau* operated together as a unit. The volume covers the design of the ship to her sinking and provides interesting, if not compelling reading. A small section of photographs and combat maps are included.

Wiper, Steve, *Kriegsmarine Scharnhorst, Warship Pictorial 36*, Classic Warships Publishing (Tucson, Arizona 2011). As the title indicates, this volume is a pictorial survey of the *Scharnhorst* in the same familiar format of this series. There are 72 pages of excellent, informative photographs with extensively annotated captions. Photographs show overall portraits of the ship as well as many detailed shots of armament, fittings and equipment. Colour photographs are included with one showing the blue of the turret crowns for the Channel Dash. It provides ideal coverage for the modeller.

WEBSITES OF INTEREST

www.bismarck-class.dk
www.feldgrau.com/kriegs.html
www.german-navy.de/kriegsmarine/index.html
www.johnrhaynes.com
www.kbismarck.com
www.kriegsmarine-reich.co.uk
www.modelwarship.com
www.Scharnhorst-class.dk
www.steelnavy.net
http://www.youtube.com/watch?v=xbgLyzoll7U (Film coverage of *Scharnhorst* & *Gneisenau*)
http://www.youtube.com/watch?v=dy0WuZhMDi0 (Still photographs of *Scharnhorst* & *Gneisenau*)

For websites of contributors, see under individual reviews.